Glacial Geology

Ice Sheets and Landforms

Second Edition

MATTHEW R. BENNETT

**School of Conservation Sciences, Bournemouth
University, UK**

NEIL F. GLASSER

**Institute of Geography and Earth Sciences
Aberystwyth University, UK**

A John Wiley & Sons, Ltd., Publication

Library of Congress Cataloging-in-Publication Data

Bennett, Matthew (Matthew R.)
    Glacial geology : ice sheets and landforms / Matthew R. Bennett,
    Neil F. Glasser. — 2nd ed.
       p.   cm.
    Includes index.
    ISBN 978-0-470-51690-4 — ISBN 978-0-470-51691-1
    1. Glacial landforms.   2. Glaciers.   I. Glasser, Neil F.   II. Title.
    GB581.B45 2009
    551.31—dc22

                                                2008052800

ISBN: 978-0-470-51690-4 (HB)
        978-0-470-51691-1 (PB)

A catalogue record for this book is available from the British Library.

Set in 10/12 Palatino by Integra Software Services Pvt. Ltd, Pondicherry India
Printed in Singapore by Fabulous Printers Pte Ltd.

First Impression   2009

Contents

Preface

In the preface for the First Edition of *Glacial Geology* we wrote that this book is the product of two things: an enthusiasm for glacial geology and a perceived need for a student text with which to stimulate this enthusiasm in others. Thirteen years on we still believe this to be the case. The First Edition has sold well and has been well-received by undergraduates studying the subject. In the First Edition of *Glacial Geology* we also wrote that the aim of the book is simple: to provide an account of glacial geology which is accessible to undergraduates and uncluttered from unnecessary detail. Although we have taken the opportunity to update and revise a lot of the content in this Second Edition, reflecting new developments in the subject, we have tried to stay true to this original aim. We hope you will still find this an accessible and helpful treatment of the subject and that by reading this book you will share some of our enthusiasm for glacial geology.

Matthew R. Bennett
Neil F. Glasser
Bournemouth and Aberystwyth, 2009

Acknowledgements

The writing of any text inevitably draws upon the accumulated wisdom of innumerable colleagues. This book is written for undergraduates and consequently it is broad in approach, with little space for the explanation of detail or to acknowledge the contribution of individuals. We have purposefully kept the text free from references as we did with the First Edition. This was a conscious student-focused choice but one which we accept does not always acknowledge the contribution of individual glaciologists and geologists. Where possible we have drawn attention to the key individuals in our choice of reading. Our initial views on glacial geology were shaped by Geoffrey Boulton and David Sugden, who not only introduced us to the subject but shared their enthusiasm and knowledge with us. Since then we have travelled to a wide range of locations and worked with many different individuals, but would like to acknowledge the contribution of the following in providing good company and challenging discussion: Mike Hambrey, Charles Warren, Richard Waller, Simon Cook, Sarita Amy Morse, Krister Jansson, Johan Kleman, Stephan Harrison, James Etienne, Bryn Hubbard, Becky Goodsell, David Graham, Nick Midgley, Sam Clemmens, Eva Sahlin, Duncan Quincey, Shaun Richardson, Brad Goodfellow, Adrian Hall, Martin Siegert, Alun Hubbard and Ted Scambos. Ian Gulley, Hillary Foxwell and Antony Smith helped draw figures for the book and the index was compiled by Sarita Amy Morse.

Illustrations

We are grateful to the following for permission to use modified or copyright figures and photographs: Allen and Unwin (Figures 4.6, 5.2, 5.3, 7.8a), American Geophysical Union (Box 3.4, Figure 4.2), Antarctic Photograph Library, US Antarctic Program (Box 6.3), Balkema (Figures 8.6, 9.20, 9.21, 9.22, 9.23, 9.24), Cambridge University Collection of Aerial Photographs (Figure 9.16), Elsevier (Figures 4.3, 6.5, 7.4, 7.11, 8.4, 8.5, 8.12, 8.14, 8.16, 9.3, 9.4, 9.6, 9.26, 9.30, 12.5, 12.6, 12.7, Boxes 1.1, 1.2, 9.2, 9.5, 12.6, 12.7), Geo Books (Figures 8.2, 8.3) The Geological Society (Figures 9.31, 10.7, 11.5, 11.6), Geological Society of America (Box 9.6), Hodder Education Group (Figures 3.1, 3.3, 3.4, 3.19, 8.21, 8.22, 10.1, 10.9, Box 3.2), INSTAAR, University of Colorado (Figure 4.4), International Glaciological Society (Figure 7.12, Boxes 5.3, 12.1), Kluwer Academic Publishers (Figures 3.6, 3.9, 3.15, 7.1), Longman (Figure 3.14), Methuen (Figures 3.10, 6.8, 6.10), NASA (Figure 2.4), National Snow and Ice Data Center (Box 2.2), Nature Publishing Group (Box 6.5), National Research Council of Canada (Figure 8.8), Norsk Polarinstittut (Figure 2.11), Quaternary Research Association (Box 8.4), Royal Geographical Society (with IBG) (Figures 3.11, 5.8), Sage Publications (Box 11.4), The Ohio State University Press (Box 7.2), UCL Press (Figures 10.6, 10.8), United States Geological Survey (Box 3.3), University of Chicago Press (Figure 8.10), Wiley-Blackwell (Tables 8.2, 8.3, Figures 3.5, 3.20, 3.22, 4.12, 6.9, 6.15, 6.16, 6.17, 8.13, 8.15, 9.1, 9.5, 9.8, 9.14, 9.15, 9.18, 9.29, 10.2, 12.1, Boxes 6.7, 8.2, 9.1, 9.8).

We are also grateful to the following people who very kindly supplied photographs and illustrations: Cliff Atkins (Box 6.9), Jonathan Bamber (Figure 2.2), Geoffrey Boulton (Figure 9.2), Jason Briner (Box 6.8), Chris Clark (Figure 12.4), Peter Doyle (Figures 10.4, 11.1), Jim Hansom (Figure 11.7), Russell Huff/Konrad Steffen (Figure 2.5), Neal Iverson (Box 5.2), Krister Jansson (Figure 2.6), Jeffrey Kargel (Figure 2.8), Martin Sharp (Figure 8.17) and Chris Stokes (Figure 12.8).

1: Introduction

1.1 WHAT IS GLACIAL GEOLOGY AND WHY IS IT IMPORTANT?

Glacial geology is the study of the landforms and sediments created by ice sheets and glaciers, both past and present. Within Earth history, ice sheets and glaciers have grown and decayed many times, making them a key part of the Earth's environmental system (Box 1.1). The present landscape in many mid-latitude areas is a function of the ice sheets and glaciers that grew and decayed during the Cenozoic Ice Age. During the Cenozoic – the past 65 million years – the Earth's climate has changed dramatically. The Antarctic Ice Sheet developed, followed by ice sheets in Greenland and elsewhere in the Arctic north. Later, large mid-latitude ice sheets developed in North America, Scandinavia, Europe, New Zealand and

---

BOX 1.1: HISTORY OF ICE ON EARTH

Ice has been part of the Earth's environmental system at several points throughout its 4.6 billion year history, and the passage of ancient ice sheets is recorded predominantly by glaciomarine sediments deposited and preserved in a variety of geological basins (Eyles, 2008). Much of this record can be interpreted by using our present understanding of glacial processes and products to interpret the past. In essence this is the application of the fundamental geological principle of uniformitarianism – the present is the key to the past. However, the Earth's Neoproterozoic glacial record challenges this idea. During the late 1990s the 'Snowball Earth' concept emerged (Hoffman and Schrag, 2002). The 'Snowball Earth' hypothesis envisages a series of cataclysmic global glaciations in which glacial ice reached tropical latitudes and the

---

Glacial Geology: Ice Sheets and Landforms Second Edition Matthew R. Bennett and Neil F. Glasser
© 2009 John Wiley & Sons, Ltd

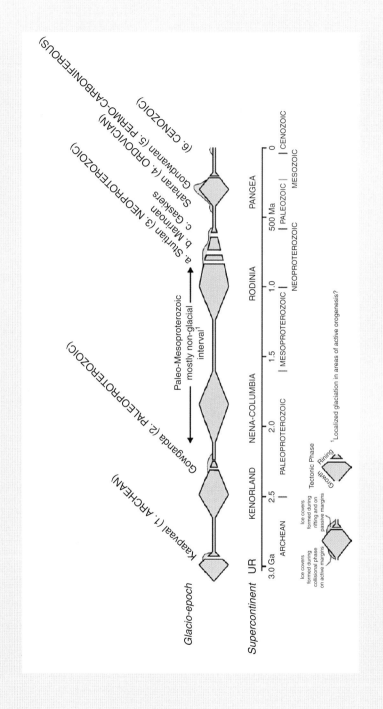

Earth's hydrological system almost completely shut down. The glacial processes involved in this global freeze would have been very different from those of the Cenozoic Ice Age. This hypothesis has become the source of considerable controversy over recent years and a number of alternative ideas have been advanced to explain the presence of glacial sediments at low latitudes at this time. One of these hypotheses, the Zipper Rift model, is based on the idea of adiabatic or uplift-driven glaciation associated with the progressive rifting of the Rodinia supercontinent. The debate surrounding 'Snowball Earth' continues to generate controversy as geologists attempt to decipher the record contained within the glacial rocks of the Neoproterozoic. The key to these debates lies in our ability to read the clues within these ancient glacial records and, in particular, in a rigorous understanding of contemporary glacial processes and products.

Sources: Eyles, N. (2008) Glacio-epochs and the supercontinental cycle after 3.0 Ga: tectonic boundary conditions for glaciation. *Palaeogeography, Palaeoclimatology, Palaeoecology*, **258**, 89–129. Hoffman, P.F. and Schrag., D.P. (2002) The Snowball Earth hypothesis: testing the limits of global change. *Terra Nova*, **14**, 129–55. Reproduced with permission from: Eyles, N. (2008) *Palaeogeography, Palaeoclimatology, Palaeoecology*, **258**, figure 1, p. 9].

BOX 1.2: CENOZOIC GLACIAL SEDIMENTS: AN ENGINEERING LEGACY

During the Cenozoic Ice Age approximately 30% of the Earth's land surface was glaciated and as a consequence over 10% of our land is now covered by glacial sediments – tills, silts, sands and gravels. In a country such as Britain this proportion is even higher. Any form of construction on or in these sediments must consider their engineering properties. At what angle will the sediment stand if excavated? How will they respond when loaded? How variable are they? How permeable are they? These questions can be answered only by a detailed knowledge of the sediments and of the processes that deposited them: the contribution of the glacial geologist.

A good example is provided by a proposed development at Hardwick Air Field in Norfolk. In 1991 Norfolk County Council applied for planning permission to build a waste hill (land raise) 10 m high to dispose of 1.5 million m^3 of domestic waste over 20 years. Crucial to their proposal was the assertion that the area was underlain by glacial till, rich in clay, which would act as a natural impermeable barrier to the poisonous fluids (leachate) generated within the decomposing waste. Normally an expensive containment

4

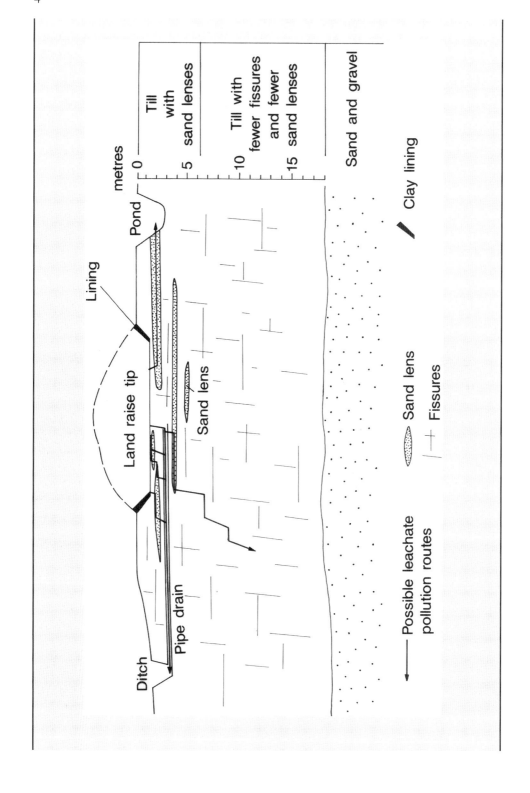

metres

0

5

10

15

Till with sand lenses

Till with fewer fissures and fewer sand lenses

Sand and gravel

Pond

Lining

Land raise tip

Sand lens

Ditch

Pipe drain

Clay lining

Sand lens

+ Fissures

Possible leachate pollution routes

liner is required to prevent contamination of the ground water by the leachate. This proposal became the subject of local debate and as a consequence the planning application was called to public planning enquiry in 1993. At this enquiry the objectors used a detailed knowledge of glacial till to argue that it was inadequate as an impermeable barrier in its natural state. Till contains fissures and pockets of sand through which the leachate may pass. The proposal was rejected, partly on the basis of this evidence. This example illustrates how knowledge of glacial sediments is vital to making engineering decisions within glaciated terrains.

Source: Gray, J.M. (1993) Quaternary geology and waste disposal in South Norfolk, England. *Quaternary Science Reviews*, **12**, 899–912. [Modified from: Gray, J.M. (1993) *Quaternary Science Reviews*, **12**, figure 9, p. 905].

Patagonia. These ice sheets dramatically changed the landscape beneath them and have left a record of their presence in the form of glacial landforms and sediments. This record shows that these ice sheets are not only a consequence of oscillations in global climate, which has driven their growth and decay with amazing regularity during the past two million years, but that they have also helped to drive climate change by modifying and interacting with the atmosphere. Understanding these ice sheets and glaciers is vital if we are to understand the mechanisms of global climate change.

In many parts of the world a distinct landscape composed of many different landforms and sediments was created by the glaciers of the Cenozoic Ice Age. This glacial landscape still survives today. It determines the distribution of valuable resources such as aggregates, and the way in which we build roads, railways, factories and houses (Box 1.2). The aesthetic appeal of this glacial landscape, to be found in many upland areas of North America and Europe, for example, is also the product of these glaciers. The spectacular mountain scenery is the result of glacial erosion, whereas glacial deposition often produces a gently rolling landscape. If we are to understand the form and texture of this glacial landscape we must understand the glaciers that produced it.

The landforms and sediments left by these glaciers are the clues from which they can be reconstructed and their behaviour studied. This subject, palaeoglaciology, is of increasing importance as we seek to understand how the glacial system interacts with other parts of the Earth's global system. By studying glacial landscapes and reconstructing the glaciers that created them we can examine the way in which glaciers grow, decay and interact with climate. From such research we can begin to predict what will happen when the mid-latitude ice sheets next return because, although the present is optimistically termed the Postglacial, there is no reason to suppose that large glaciers or ice sheets will not return to the mid-latitudes in the future.

1.2 THE AIM AND STRUCTURE OF THIS BOOK

Glaciers are the scribes of the Cenozoic Ice Age and they have etched its story upon the landscape. It is a story that has been written repeatedly upon the same page with the successive growth and decay of each glacier. Each glacier has destroyed, remoulded or buried the evidence of earlier phases of glaciation. Deciphering the story of this complex geological record is therefore difficult and requires careful detective work. The landforms and sediments left by former glaciers provide the clues from which to reconstruct their form, mechanics and history. This book shows you how to interpret these clues.

We start first by looking at contemporary glacial environments around the world in order to illustrate the diversity of the glacial systems that exist today. This is followed by two chapters that introduce the basic mechanics of the glacial system to provide an understanding of how glaciers work. We explain how ice sheets and glaciers grow, flow and decay. In Chapters 5 and 6 we explore the processes of glacial erosion and consider the landforms that they create; landforms which can be seen in the landscape today and which provide information about the dynamics of the glaciers that created them. In Chapters 7–11 we tackle the processes of glacial sedimentation and landform development, all of which provide important evidence of glacier activity. The final chapter examines how we can use the clues in the landscape to reconstruct ancient glacial systems – the study of palaeoglaciology. Important terms in the text are highlighted in *italics* either when they are first used or when they are particularly pertinent to the subject being considered. Some terms therefore will appear in *italics* more than once.

2: Glaciations Around The Globe

The aim of this chapter is to illustrate, via series of case studies, the range and diversity of styles of glaciation on Earth today, while also drawing attention to some of the contemporary debates within the discipline that are focused on these different regions (Figure 2.1). Those readers without any knowledge of glacial processes may choose to skip this chapter and return to it later. The first two case studies describe the contemporary polar ice sheets, and cover: (i) the recent changes in the glaciers and ice shelves around the Antarctic Ice Sheet; and (ii) the complex relationship between changes in outlet glacier discharge and climate on the Greenland Ice Sheet. The remaining five case studies are concerned with the styles of glaciation in contrasting glaciological settings, and include: (i) southern hemisphere temperate glaciers in Patagonia and New Zealand; (ii) northern hemisphere temperate glaciers in Alaska and Iceland; (iii) high-altitude glaciers in the Himalaya; (iv) tropical glaciers in the Cordillera Blanca, Peru; and (v) polythermal glaciers in the Arctic.

2.1 THE ANTARCTIC ICE SHEET

The Antarctic Ice Sheet, covering 98% of the continent, is the largest ice sheet on Earth. The ice sheet averages ~1.6 km thick, but it is over 4 km thick where it overlies deep subglacial basins. It is divided in two by the Transantarctic Mountains, with the smaller West Antarctic Ice Sheet on one side and the larger East Antarctic Ice Sheet on the other. The continent contains about 90% of the world's glacier ice, which equates to about 70% of the entire world's fresh water. If the ice sheet melted, sea levels would rise globally by about 60 m. In most of the interior of the continent precipitation is very low, often as little as 20 mm per year, although precipitation rates rise towards the coast, where the air contains more moisture. Of the two components of the ice sheet, the West Antarctic Ice Sheet has received most scientific attention because of the possibility that it could collapse, or disintegrate rapidly. The reason for this potential instability stems from the fact that the ice sheet overlies a basin with a mean elevation below contemporary sea level. If the West Antarctic Ice Sheet were to collapse, global sea levels could rise by up to 6 m in a matter of centuries. As a whole the Antarctic Ice Sheet is a complex system, but we can identify six main components.

Glacial Geology: Ice Sheets and Landforms   Second Edition   Matthew R. Bennett and Neil F. Glasser
© 2009 John Wiley & Sons, Ltd

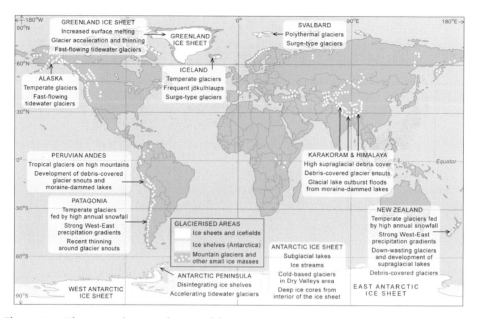

Figure 2.1 The main glacierised areas of the world showing the locations of the case studies presented in this chapter and some of the key attributes of glaciers in these areas.

1. **A high-elevation plateau or 'Polar plateau'.** Here there is little moisture and so snow accumulates very slowly such that it is measured in millimetres to centimetres per year. It has been suggested that peripheral thinning of the ice sheet, which has been observed recently, is balanced by interior thickening on the Polar plateau, but the absence of detailed mass balance studies of the ice sheet makes this difficult to establish with any degree of certainty. As a result, we do not know if the ice sheet is in negative or positive mass balance (see Section 3.1). The longest continuous deep ice cores have been drilled on the Polar plateau, providing information about the climatic and atmospheric records in Antarctica over eight glacial cycles spanning the past 740 000 years.

2. **Peripheral ice streams.** These ice streams are key components of the glacial system because they discharge the majority of the ice and sediment associated with the Antarctic Ice Sheet; for example, ice streams account for over 90% of the overall discharge from the ice sheet (Figure 2.2). Ice streams are narrow corridors of fast flowing ice with velocities in the range of 0.5–1 km per year, over two orders of magnitude faster than adjacent ice. Consequently, ice streams are typically heavily crevassed, with abrupt lateral margins (*shear margins*) between the ice streams and surrounding ice. The variables that control the location, dynamics and behaviour of these ice streams are uncertain, but in some cases they appear to overlie sedimentary basins of soft-sediment that may deform beneath the ice thereby assisting forward flow (see Section 3.3). To add to this complexity it is now known that ice streams can switch on or switch off through

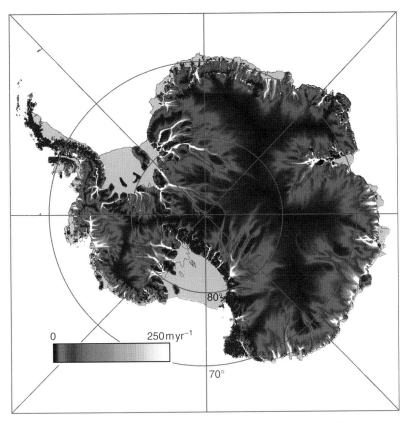

Figure 2.2   Balance velocity map of the Antarctic Ice Sheet. Note how the ice sheet is generally slow-flowing in the interior and is drained by a number of fast-flowing outlet glaciers and ice streams around the periphery. [Image courtesy of: Jonathan Bamber]

time. For example, the Kamb Ice Stream, formerly known as Ice Stream C, in West Antarctica switched off around 150 years ago and has been stagnant ever since.

3. **Fringing ice shelves.** These are the floating continuation of outlet glaciers and surround nearly half of the Antarctic continent. Some of these ice shelves are very large, such as the Ross Ice Shelf, which is about the same area as France. Ice shelves perform a vital role in influencing the dynamics, and therefore the system response time, of upstream ice from inland Antarctica because they hold back or buttress ice flow from the glaciers or ice streams that feed them (Box 2.1). Calving of icebergs from ice-shelf termini accounts for approximately 90% of Antarctic ice loss. There is also rapid heat exchange between the ice shelves and the ocean beneath, meaning that they add considerable quantities of fresh water to the oceans. Finally, ice shelves are capable of entraining, transporting and depositing large quantities of glacial sediment (Figure 2.3) and dispersing it more widely as calved icebergs drift out to sea.

BOX 2.1: ANTARCTIC PENINSULA ICE SHELVES: BREAKING UP IS ALL TOO EASY

Some of the most dramatic recent changes in Antarctica have taken place on the Antarctic Peninsula, the northernmost part of the continent that projects towards South America. As a direct result of temperature rise in the past 50 years or so, ice shelves on the Antarctic Peninsula have been slowly receding. Most of the ice shelves have shown two phases of recession; a climatically driven progressive recession over decades, and a more rapid *ice-shelf collapse* phase. Probably the most famous event was the sudden collapse over 6 weeks (February–March 2002) of the Larsen B Ice Shelf, which disintegrated with the loss of 3200 km^2 of ice shelf. The recent mass loss from glaciers formerly feeding the ice shelves of the Antarctic Peninsula is estimated to be sufficient to raise eustatic sea level by between 0.1 and 0.16 $\pm$ 0.06 mm yr^{-1}.

The 2002 Larsen B Ice Shelf collapse has been studied intensively using a number of methods. The key features of the ice-shelf collapse are:

1. Prior to its 2002 collapse, the Larsen B Ice Shelf had been in existence throughout the entire Holocene (the past 10 000 years). The presence or absence of ice shelves can be detected in the sediments around the ice shelves using ocean cores (Domack *et al.*, 2005).
2. Before it collapsed the ice shelf was fed by a number of individual tributary glaciers originating from the mountains of the Antarctic Peninsula.
3. By 2002, the ice shelf had thinned to a critical level following decades of warming, possibly to a point where structural glaciological weaknesses in the ice shelf such as crevasses and rifts caused rapid ice-shelf collapse (Glasser and Scambos, 2008).
4. Large meltwater ponds appeared on the ice-shelf surface in the summers prior to its collapse in 2002. It has been suggested that this meltwater acted as a mechanical force in the crevasses causing breaks in the ice shelf and thus accelerating ice-shelf disintegration (MacAyeal *et al.*, 2003).
5. After 2002, the tributary glaciers responded to ice-shelf collapse by rapid thinning and acceleration (Scambos *et al.*, 2004). This illustrates the buttressing effect of the shelf.

The images below show the collapse of the Larsen B Ice Shelf between January and March 2002 as recorded by NASA's MODIS satellite sensor. The images show the Larsen B Ice Shelf and parts of the Antarctic Peninsula (on the left). The first scene from 31 January 2002 shows the shelf in late austral summer, with dark melt ponds dotting its surface. In the next two scenes minor retreat takes place, amounting to about 800 km^2, during which time several of the melt ponds well away from the ice-front drained through new cracks within the shelf. The main collapse is seen in the last two scenes, on 5 March and 7 March, with thousands of sliver icebergs and a large area of very finely divided 'bergy bits' where the shelf formerly lay. The last phases of the retreat totalled ~2600 km^2. Resolution of the original images is 500 m.

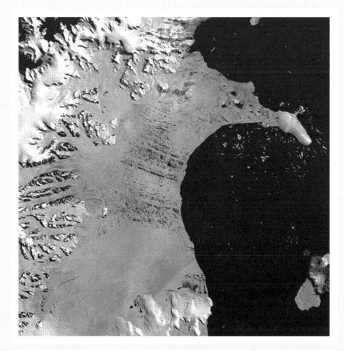

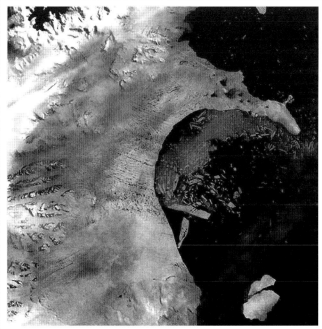

Sequence of four MODIS satellite images showing the disintegration of the Larsen B Ice Shelf in February–March 2002.

Sources: Domack, E., Duran, D., Leventer, A., *et al.* (2005) Stability of the Larsen B ice shelf on the Antarctic Peninsula during the Holocene epoch. *Nature*, **436**, 681–5. Glasser, N.F. and Scambos, T.A. (2008). A structural glaciological analysis of the 2002 Larsen B ice shelf collapse. *Journal of Glaciology*, **54**, 3–16. MacAyeal, D.R., Scambos, T.A., Hulbe, C.L., *et al.* (2003). Catastrophic ice-shelf break-up by an ice-shelf-fragment-capsize mechanism. *Journal of Glaciology*, **49**, 22–36. Scambos, T.A., Bohlander, J.A., Shuman, C.A. and Skvarca, P. (2004). Glacier acceleration and thinning after ice shelf collapse in the Larsen B embayment, Antarctica. *Geophysical Research Letters*, **31**, L18402. [Image courtesy of: NSIDC]

4. **Tidewater glaciers.** Where glaciers reach sea level but do not spread out to form an ice shelf they terminate as tidewater glaciers, and calve rapidly to provide another efficient means of ice loss from the ice sheet (Figure 2.4). Some glaciers, such as the neighbouring Thwaites Glacier and Pine Island Glacier, have been observed to be accelerating rapidly in recent years, at rates of up to 5% per year faster, leading to concerns that they are undergoing rapid and irreversible

Figure 2.3   The surface of the McMurdo Ice Shelf in Antarctica showing extensive debris cover and the development of surface meltponds. [Photograph: N.F. Glasser]

changes to their velocity structures. This is of concern because if this trend continues then the draw-down of ice from inland Antarctica could de-stabilise the ice sheet and potentially raise global sea levels rapidly.

Figure 2.4 Part of an ASTER satellite image acquired on 12 December 2000 showing the calving front of Pine Island Glacier in Antarctica. The image covers an area of 38 × 48 km. Note the large incipient fracture approximately 20 km behind the calving front. [Image courtesy of: NASA/GSFC/METI/ERSDAC/JAROS, and US/Japan ASTER Science Team]

5. **Subglacial lakes.** These are large freshwater reservoirs deep beneath the ice sheet and are found in areas of basal melting, due to high geothermal heat flux, that overlie topographic depressions. More than 70 subglacial lakes have now been identified beneath the Antarctic Ice Sheet, including the largest, Lake Vostok, which lies beneath 3 km of ice and is 230 km long, 14 000 km^2 in area, with a water volume of around 2000 km^3. Satellite observations of the ice surface elevation in the vicinity of the subglacial lakes indicate that the lakes are connected to one another subglacially and that subglacial water moves rapidly between individual lakes by drainage beneath the ice sheet. There is landform evidence in the shape of large meltwater channel systems that some of the former Antarctic subglacial lakes may have drained catastrophically in the past (see Box 6.3).

6. **Outlet glaciers and valley glaciers.** Away from the fast-flowing ice streams are smaller glaciers. The most famous of these, because they feed into areas of ice-free

land, are those of the Dry Valleys (Antartica). These cold-based glaciers have remarkably low accumulation rates of only a few centimetres per year, very low surface temperatures ($\sim -30°C$), very low basal temperatures ($\sim -15°C$) and little or no meltwater is present within the ice. As a result they have very low ice velocities and consequently documented rates of glacial erosion and landscape evolution are remarkably low in this area. In fact rates of landscape modification are so low that this environment contains a record of landscape change that can be measured in millions of years. Cold-based glaciers commonly have substantial basal debris loads, especially near their margins where stacking of debris sequences due to regelation is common and significant thicknesses of debris-rich ice can be generated, with basal debris layers of up to 5 m thickness.

2.2 GREENLAND IN THE GREENHOUSE

The Greenland Ice Sheet is the second largest ice sheet on Earth and is considered to be especially vulnerable to global warming. There are two reasons for this: first it spans a range of latitudes from 60° to 83°N; and second it is surrounded by ocean waters, which unlike Antarctica do not possess ocean currents that isolate the ice sheet from mid-latitude heat transfer. Studies show that the ice sheet has experienced record amounts of surface melting in recent years (Figure 2.5). Many of the outlet glaciers that drain the Greenland Ice Sheet have also accelerated and thinned

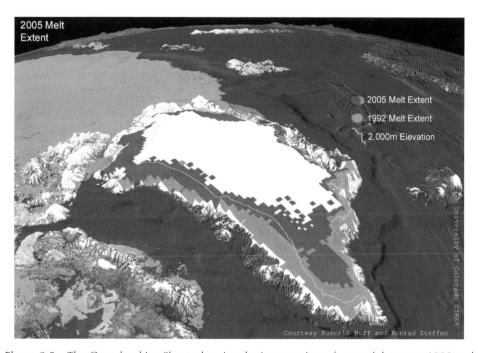

Figure 2.5   The Greenland Ice Sheet, showing the increase in surface melt between 1992 and 2005. [Image courtesy of: Russell Huff and Konrad Steffen]

in the past decade. The net loss in volume of the Greenland Ice Sheet has more than doubled in recent years, from 90 to 220 km^3 per year. The main cause of this increase is the acceleration of several large outlet glaciers. Three outlet glaciers in particular, Jakobshavn Isbrae, Helheim Glacier and Kangerdlugssuaq Glacier, have seen an exceptional speed-up, surface lowering and recession. These changes are important because, between them, these three glaciers drain around 12% of the ice from the Greenland Ice Sheet. The glaciers could potentially draw-down large volumes of ice from the interior of the ice sheet and increased ice discharge to the oceans would contribute directly to sea-level rise.

Glaciologists are trying to understand the mechanisms behind these recent changes using both field studies and remotely sensed data. It is widely accepted that the observed thinning of the glaciers results from changes in the flow regime of the glaciers rather than changes in surface mass balance, but the reasons for the initiation of these dynamic changes remains to be fully explained. Two possible processes have been suggested.

1. The first mechanism is the *Zwally effect*, which relies on meltwater reaching the bed of the ice sheet and reducing friction through a higher basal water pressure. This follows from the observation that increased air temperatures in the region in the past few decades have led to longer or more widespread surface melting. The mechanism is simple: warmer air temperatures mean increased surface melting, so there is more water on the ice-sheet surface. As this surface meltwater drains through moulins, crevasses or other fractures to the glacier bed, it provides lubrication and reduces friction so that the glacier velocity due to sliding increases. This process, originally proposed by Jay Zwally, was observed to be the cause of a brief seasonal acceleration of up to 20% on the Jakobshavns Glacier in 1998 and 1999. The acceleration lasted two to three months. Zwally hypothesised that the coupling between surface melting and ice-sheet flow provides a mechanism for rapid, large-scale, dynamic responses of ice sheets to climate warming. More recent studies also show that rapid supraglacial lake drainage events are related to short-term velocity fluctuations. However, it is unclear if these events are on a sufficient scale to have a lasting or significant effect on the annual flow of large outlet glaciers.

2. The second mechanism is the *Jakobshavn effect*, a term coined by Terry Hughes to describe a situation where a small imbalance of forces in a calving glacier causes a substantial non-linear response in its behaviour. In this case the imbalance of forces is at the calving front and propagates up-glacier. For example, thinning at the calving front might cause the glacier to become more buoyant, or to float at the snout, making it more responsive to tidal changes. Reduced friction due to greater buoyancy causes the rate of iceberg calving to increase, leading to glacier acceleration and draw-down. This process has been invoked to explain the exceptional thinning of the three outlet glaciers. Other possible triggers for perturbations in tidewater glacier behaviour are changing marine influences, such as warmer waters and the decreasing influence of sea-ice in the fjords, or decreases in water depth due to sedimentation.

Understanding the relative importance of these two mechanisms and their link to climate change is critical if we are to make century-scale predictions about future changes in the Greenland Ice Sheet and its contribution to sea-level rise. This is because the second mechanism (the 'Jakobshavn effect') will cease to operate once the ice sheet retreats beyond the influence of the ocean, whereas the first mechanism (the 'Zwally effect') will not. Calving-induced dynamic change is therefore self-limiting, whereas meltwater-drainage-induced dynamic change is not.

One major problem with the meltwater-drainage hypothesis is the assumption that an increase in subglacial meltwater availability automatically leads to increased ice flow, partly because our observations of this effect on the Greenland Ice Sheet cover only a few years. Also, observations on smaller valley glaciers suggest that increased ice flow is associated only with the initial (e.g., 'spring event'; see Section 4.7) phase of meltwater drainage to the bed, and that later phases, although characterised by far greater amounts of meltwater, are not associated with rapid flow. This is because the subglacial drainage system changes its configuration to accommodate the increased water flux. If the Zwally effect is the key then, because meltwater is a seasonal input, velocity will have a seasonal signal. If the Jakobshavn effect is the key, the velocity will propagate up-glacier and there will be no seasonal cycle. This is clearly testable given good real-time observations on glacier velocity. It should also be remembered that just as ice discharge can increase suddenly, so it can decrease suddenly: Helheim and Kangerdlugssuaq Glaciers doubled their discharge within a year in 2004, but 2 years later this discharge had quickly dropped back to close to its former rate. It is therefore important to note that so far no one has actually shown that increased surface meltwater is responsible for the acceleration of the outlet glaciers.

2.3 SOUTHERN HEMISPHERE TEMPERATE GLACIERS: PATAGONIA AND NEW ZEALAND

Numerous ice caps, outlet glaciers and independent valley glaciers occur in the mountainous areas of southern South America (Patagonia) and in New Zealand. In Patagonia there are many glaciers in the Andes south of 46°S, fed by high precipitation from the Southern Westerlies (Box 2.2). Annual precipitation on the large icefields here is estimated to be as high as 10 000 mm water equivalent – one of the highest in the world. As a result, the western glaciers have high mass-balance gradients and are very dynamic. Strong west to east precipitation gradients mean that the eastern glaciers are less dynamic (Figure 2.6).

In New Zealand the main area of glacierisation is in the Southern Alps between 43 and 45°S. On the western side of the Main Divide, in a maritime climate, annual precipitation is high. The terminus of the Franz Josef Glacier, for example receives over 5000 mm of precipitation per year, increasing to ~15 000 mm in the névé above the equilibrium line. Consequently the glaciers are fast-flowing with measured centre-line velocities of 600 m per year at the Franz Josef Glacier. Glaciers to the east of the Main Divide lie in a much drier climate and there is an associated steep eastward rise of glacier equilibrium line altitudes (Figure 2.7). There are many striking similarities between the characteristics of the glaciers in Patagonia and New Zealand.

BOX 2.2: THE PATAGONIAN ICEFIELDS

There are many glaciers in the Andes south of 46°S in Patagonia. Two major ice masses exist in the region. First is the ~4200 km^2 North Patagonian Icefield (47°00'S, 73°39'W), which is some 120 km long and 40–60 km wide, capping the Andes between altitudes of 700–2500 m a.s.l. Annual precipitation on the western side of the icefield increases from 3700 mm at sea level to an estimated maximum of 6700 mm at 700 m a.s.l. Precipitation decreases sharply on the eastern side of the icefield. The North Patagonian Icefield contains the lowest latitude marine-terminating glacier on Earth, the San Rafael Glacier, which descends through rainforest to calve icebergs into a coastal Laguna as illustrated below. Second is the larger (~13 000 km^2) South Patagonian Icefield, which stretches north to south for 360 km between 48°50' and 51°30'S with a mean width of ~40 km. The southern icefield again has very strong west to east precipitation contrasts, as well as some very large tidewater and lake-calving glaciers (Aniya, 1999). There are also many smaller satellite glaciers on the mountains surrounding the two icefields. There is strong evidence that the glaciers of the two icefields are very sensitive to recent climate change. As a result, they have been receding and thinning (Rivera *et al.*, 2007). The glaciers of the two icefields have therefore contributed to recent sea-level rise. Rignot *et al.* (2003) estimated that the two icefields jointly contributed 0.042 ± 0.002 mm yr^{-1} to global sea level in the period 1968/1975 to 2000 but that this has doubled to 0.105 ± 0.011 mm yr^{-1} in the more recent years 1995–2000.

Sources: Aniya, M. (1999) Recent glacier variations of the Hielos Patagonicos, South America, and their contribution to sea-level change. *Arctic, Antarctic and Alpine Research*, **31**, 165–73. Rignot, E., Rivera, A. and Casassa, G. (2003) Contribution of the Patagonia Icefields of South America to sea level rise. *Science*, **302**, 434–7.

Rivera, A., Benham, T., Casassa, G. *et al.* (2007) Ice elevation and areal changes of glaciers from the Northern Patagonia Icefield, Chile. *Global and Planetary Change*, **59**, 126–37. [Photograph: N.F. Glasser]

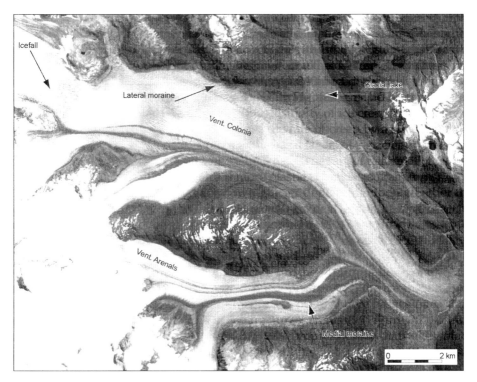

Figure 2.6   ASTER satellite image of Glaciers Colonia and Arenales on the eastern side of the North Patagonian Icefield. Note the increase in surface debris cover down-glacier as a result of surface ablation. [Image courtesy of: Krister Jansson]

1. In both cases the glaciers are fed by large amounts of precipitation that falls as snow at higher elevations from southwesterly weather systems, with very strong precipitation gradients between the west (maritime) and the east (continental).
2. The glaciers have steep *mass balance gradients* (see Section 3.2), with extremely high rates of snow accumulation and high ablation rates.
3. Both regions are dominated by temperate ('warm') glaciers, where the ice at the bed is at the pressure-melting point throughout.
4. Outlet glaciers terminate in a wide variety of environments, with examples of land-terminating glaciers (both Patagonia and New Zealand), lake-calving glaciers (both Patagonia and New Zealand) and marine-terminating glaciers (Patagonia).

5. Glaciers in both regions have been receding and thinning dramatically in recent years. The volume of ice in the Southern Alps of New Zealand has reduced by about 5.8 km^3, or almost 11%, in the past 30 years. More than 90% of this loss is from 12 of the largest glaciers and is inferred to be a direct response to rising temperatures over the twentieth century.

6. Most of the mass loss has been in the terminal sections of the glaciers, where rates of down-wasting are measured in metres per year. Glaciers to the east of the Main Divide in New Zealand and to the east of the Andes in Patagonia are bordered by 100 m high lateral moraines formed as the glaciers down-waste.

7. As the glaciers down-waste, large proglacial melt lakes form around their termini, for example the Tasman and Hooker Lakes in New Zealand. Large areas of the glaciers may begin to float where they terminate in these lakes, promoting rapid subaqueous melting and further mass loss due to iceberg calving.

Figure 2.7   The snout of the Tasman Glacier in New Zealand, showing the extent of recent (twentieth and twenty-first centuries) down-wasting. Note the debris-covered snout and large lateral moraines, which impound the proglacial lake. [Photograph: N.F. Glasser]

There are also similarities in the typical subglacial processes operating beneath these glaciers and the landforms and sediments that they produce.

1. The presence of basal meltwater means that glaciers in both areas move by *glacier sliding*, *subsole deformation* or a combination of both (see Section 3.3). As a result, they

carry little debris at the bed. Basal debris-rich layers are thin, of the order of centimetres to tens of centimetres, because of the substantial volumes of meltwater flowing there. These glaciers are also fast-flowing and slide over their beds, and are therefore normally regarded as efficient agents of glacial erosion.

2. The glacier surfaces are commonly debris-mantled, with extensive areas of *supraglacial debris* derived from rockfall activity on surrounding valley walls and slopes. This supraglacial debris is either organised into *medial* and *lateral moraines*, both of which are important features of high-level sediment transport, or forms a continuous cover on the ice surface. *Outburst floods* (jökulhlaups) from englacial or subglacial lake drainage can also reach the glacier surface to contribute to the supraglacial debris load.

3. Glacial landforms and sediments are varied, reflecting the complexity of the sediment transport and depositional processes operating. Supraglacial debris supply and sedimentation are important, particularly where the debris becomes concentrated on the ice surface during down-wasting. A recent study of New Zealand glaciers highlights the importance of *glaciofluvial* sediment transfer, suggesting that the dominant sediments around the glaciers are glaciofluvial, transported by supraglacial, subglacial and proglacial streams. Reworking, particularly by proglacial streams, dominates the final depositional products. Consequently many of the 'glacial' sediments are actually glaciofluvial and therefore lack a distinct glacial imprint.

2.4 NORTHERN HEMISPHERE TEMPERATE GLACIERS: ALASKA AND ICELAND

Temperate glaciers are found in the maritime mountains of Alaska and in Iceland. The high rates of snowfall and ablation in Alaska give rise to some of the steepest mass balance gradients in the world (see Section 3.2) and consequently some of the fastest glaciers, with ice velocities in some cases up to 500 m per year. The steep mountains surrounding many Alaskan glaciers give these glaciers large supraglacial debris components and complex ice margins with extensive spreads of supraglacial sedimentation. The high rates of ablation mean that meltwater is abundant at their ice margins and consequently the glaciers are often well connected to large proglacial fluvial systems. Glacial debris is released efficiently to proglacial rivers and consequently extensive areas of ice-marginal landforms do not develop. The other key feature of Alaska is the concentration of glaciers showing *surge-type behaviour*. Globally, the majority of glaciers have a simple relationship between the mass balance gradient and glacier velocity, in which a glacier flows through the year in order to move snow accumulated in the upper reaches of the glacier to the snout where ablation is dominant. However, a small proportion of glaciers (~10%) behave differently and undergo prolonged periods of quiescence when the ice flows slowly, punctuated by periods of extremely rapid flow that last a few months or years. In essence the glacier is subject to a 'binge–purge' style of flow behaviour. Surge-type glaciers are often associated with rapid ice-frontal advance. Understanding the mechanics and thresholds involved in this process is an area of active research, concentrating on some form of instability within the flow mechanics of a particular glacier (see Section 3.5). Glaciers such as Variegated and Black Rapids Glaciers in Alaska have been studied extensively to yield important information on the cause of surge-type behaviour, although knowledge of these processes remains incomplete.

Another Alaskan glacier that has been studied extensively is the Matanuska Glacier, which is especially associated with studies of the sedimentology of *flow tills* (see Section 8.1.2). More recently, this glacier has been at the centre of a controversy concerning the role of *glaciohydraulically supercooled meltwater* in debris entrainment and transfer (see Section 7.2). Thick layers of debris-rich ice occur along the margins of the glacier. These have been interpreted as the result of the freezing-on of debris within supercooled meltwater rising from an overdeepened basin close to the ice margin. The freezing point of ice is pressure-dependent and water may remain liquid below 0°C under thick ice, for example at the base of an overdeepened basin. As subglacial meltwater ascends the adverse slope towards the margin to escape the overdeepening, the pressure-dependent melting point rises as the overburden pressure falls. If the reverse slope is sufficiently steep, the water cannot warm sufficiently quickly so it emerges in a supercooled state, leading to the build up of thick layers of debris-rich basal ice as it freezes on to the glacier sole. This process has been proposed as a means of creating thick basal ice layers, but it remains controversial and it is not universally supported.

Iceland also provides a good example of an area dominated by temperate glaciers, and over the past 30 years this region has consistently been a focus for glacial geology research. In particular the southern outlet glaciers of the Vatnajökull Ice Cap have been studied extensively, leading to many important conceptual advances. For example, the concept of *subglacial deformation* (see Box 3.4) – the idea that a glacier may flow by deformation of the sediment beneath it – originated in part from direct field observations made at Breiðamerkurjökull. Some of the outlet glaciers of Vatnajökull surge at periodic intervals and have contributed to our understanding of the landform assemblages associated with surge-type behaviour, for example the crevasse-squeeze ridges that occur behind large push moraines. At the non-surge type glacier margins three depositional domains are common: (i) areas of extensive, low-amplitude marginal, dump, push and squeeze moraines; (ii) incised and terraced glaciofluvial forms; and (iii) subglacial landform assemblages of flutes, drumlins and overridden push moraines.

2.5 HIGH-ALTITUDE GLACIERS: THE HIMALAYA

The Himalaya (Figure 2.8) extends for ~2000 km from Afghanistan to Burma. Outside the polar regions it is one of the most extensively glacierised regions of the world, but one that is poorly known from a glaciological perspective. The Himalaya is influenced by two major weather systems, the South Asian monsoon and the mid-latitude westerlies. Probably the most studied glaciers are those in the Everest region, where the dominant influence is the South Asian monsoon from the Indian Ocean, which results in a pronounced summer maximum of precipitation

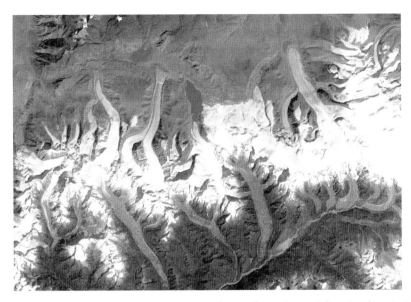

Figure 2.8   ASTER satellite image of glaciers on the northern slope of the Himalaya in Bhutan. There is evidence that the glaciers are receding in the form of well-developed terminal and lateral moraines, as well as the development of proglacial moraine-dammed lakes [Image courtesy of: Jeffrey Kargel, USGS/NASA JPL/AGU].

from late-May to mid-September. Maximum precipitation occurs at altitudes between 5000 and 7000 m. Although average daily temperatures at this altitude are below 0°C for much of the year, high winds prevent extensive accumulation of snow at high altitudes. Unlike polar or lower-altitude alpine glaciers, the coincidence of the highest temperatures and maximum precipitation means that the peak of accumulation and ablation occur simultaneously. This glaciologically unusual regime profoundly influences the manner in which debris is transported and entrained in these glaciers.

There are of course local variations in glacier type within the Himalaya, but it is possible to make the following generalisations.

1. The steep terrain in the high-altitude catchments mean that accumulation is commonly dominated by ice and snow avalanching, both of which incorporate considerable quantities of frost-shattered rock. The glaciers therefore commonly carry a heavy supraglacial debris load derived from this rockfall material. Much of the rock material is incorporated englacially and in the ablation zone it becomes concentrated to form a near-continuous debris mantle. Texturally this material is predominantly a sandy boulder-gravel, but with minor amounts of boulder-gravel and diamict.
2. Many of the glaciers have prominent (up to 100 m high) lateral and frontal moraines, dating from the Little Ice Age or earlier. The moraines often have

vegetated outer faces and an unvegetated, loose, collapsing inner face. In extreme cases, for example on the Khumbu Glacier in Nepal, *terminal moraines* may reach height of 250 m above the surrounding land. Geophysical surveys indicate that these moraines have a substantial core of dead ice within them. Terminal and lateral moraines are composed mainly of sandy boulder-gravel derived from a mixture of rockfall debris and material from the zone of traction along the valley sides.

3. Where they are active, measured surface velocities on the largest Himalayan glaciers are between 10 and 100 m per year. Many Himalayan glaciers, however, have much slower flow rates especially in their terminal areas. Therefore, rather than receding in an active manner, many glaciers are undergoing down-wasting and therefore becoming debris-mantled in their lower reaches. The debris cover is mostly hummocky, and dominated by coarse angular material. Between the debris hummocks are depressions containing meltwater ponds, with steep exposed ice slopes smeared with debris, or ice cliffs several metres high. The thicker cover of debris towards the terminus retards ablation more than the progressively thinner debris cover up-glacier, resulting in a lessening or even reversal of the surface gradient. When the overall average gradient declines to less than 2°, large *supraglacial ponds* begin to form on the glacier snout. Over time, these lakes coalesce to form large *supraglacial lakes*, dammed by the terminal and lateral moraines. Where they are impounded by ice-cored moraines, these lakes form a natural hazard because the moraine dams are prone to sudden failure, resulting in *glacial lake outburst floods* (GLOFs; Box 2.3).

4. Glacial rivers in the Himalaya often lack extensive braided reaches in their proximal zones. In comparison with alpine regions therefore, proximal glacio-fluvial processes are relatively unimportant in these high-altitude environments. Meltwater is largely confined to single channels because of the steepness of the channel gradient and narrowness of the valleys.

5. Below many of the glaciers are breached terminal moraines and well developed alluvial fans representing the tracks of former GLOFs (Box 2.3). Peak GLOF discharges have been calculated as up to 60 times greater than seasonal high flow floods derived from snowmelt runoff, glacier meltwater and monsoonal precipitation. As a result, these floods are tremendously powerful, eroding,

BOX 2.3: GLACIER HAZARDS IN THE HIMALAYA AND THE ANDES

As glaciers in mountain areas recede in response to climatic warming, a number of different *glacier hazards* can develop. Richardson and Reynolds (2000) have identified two main types of glacial hazard (Table 1). *Direct glacial hazards* involve the direct action of snow and ice and include events such as snow and ice avalanches, glacier outburst floods and glacial advances. *Indirect glacial hazards* arise as a secondary consequence of a glacial feature or process

Table 1 Types of glacier, glacial and related hazards

Category	Hazard event	Description	Time scale of event
Direct	Avalanche	Slide or fall of large mass of snow, ice and rock	Minutes
	Glacier outburst flood	Catastrophic discharge of water under pressure from a glacier	Hours
	Jökulhlaup	Glacier outburst flood often associated with subglacial volcanic activity	Hours to days
	Glacier surge	Rapid increase in rate of glacier flow	Months to years
	Glacier fluctuations	Variations in ice-front positions	Years to decades
Indirect	Glacial lake outburst floods (GLOFs)	Catastrophic outburst, typically from a moraine-dammed proglacial lake	Hours
	Débâcle	French term for an outburst from a proglacial lake	Hours
	Aluvión	Spanish term for a catastrophic flood of liquid mud	Hours
	Lahars	Catastrophic debris flow associated with rapid melt during volcanic activity	Hours
	Water resource problems	Water supply shortages, particularly during low flow conditions, associated with wasting glaciers and climate change, etc.	Decades

and may include catastrophic breaching of moraine-dammed lakes or water resource problems associated with wasting glaciers and climate change. One of the principal glacier hazards is the threat posed by catastrophic drainage of hazardous moraine-dammed lakes, often known as glacial lake outburst floods (GLOFs). The number and volume of potentially hazardous moraine-dammed lakes is increasing in both the Himalaya and the Andes. These lakes develop behind unstable ice-cored moraines, and have the potential to burst catastrophically, producing devastating floods. Discharge rates of $30\,000\,\text{m}^3\,\text{s}^{-1}$ and run-out distances in excess of 200 km have been recorded from Himalayan GLOFs. Glacial hazards have attracted attention recently for two main reasons: (i) the risk of loss of life; and (ii) the serious threat to costly infrastructures such as hydropower installations, roads and communications. Carey (2005) has discussed the impacts of these glacier hazards on people living with these risks in Peru. The image below shows the aftermath of a glacial lake outburst flood (GLOF) in the Nepal Himalaya showing a large breach through a moraine and spread of gravel deposited during the GLOF.

Source: Carey, M. (2005) Living and dying with glaciers: people's historical vulnerability to avalanches and outburst floods in Peru. *Global and Planetary Change*, **47**, 122–34. Table modified from: Richardson, S.D. and Reynolds, J.M. (2000) An overview of glacial hazards in the Himalayas. *Quaternary International*, **65/66**, 31–47. [Photograph: N.F. Glasser]

transporting and depositing large quantities of sediment for tens of kilometres along their flood routes.

2.6 TROPICAL GLACIERS: THE CORDILLERA BLANCA, PERU

We normally think of glaciers as occupying only cold places so it may come as a bit of a surprise to learn that *tropical glaciers* occur near the Equator on the mountains of East Africa, Indonesia and in the Andes. On a global scale, the total area of these tropical glaciers is small ($\sim$2500 km^2) and they make up only 4% of the area of all

Earth's mountain glaciers which is equivalent to about 0.15% of the Earth's total glacier area. More than 99% of all tropical glacier area is in the South America Andes between Bolivia and Venezuela, with more than 70% in the Peruvian Cordilleras alone. The tropical climate is very different to the mid-latitude or polar environments so glaciers can exist at these low latitudes only where there are mountains of sufficient altitude to allow snow accumulation. Climatically, these mountains are characterised by a homogeneous atmosphere without frontal activity, a lack of thermal seasonality and by very pronounced precipitation seasons. Around the tropical glaciers of the South American Andes, for example, the year can be divided into three climatic periods: the dry season (May–August), the wet season (January–April) and a transition period (September–December), where there is a gradual build up in precipitation towards the wet season. As a result of this our normal concepts of simple accumulation and ablation seasons (see Section 3.1) do not hold for tropical glaciers. In many locations the period of maximum precipitation, and therefore snow accumulation, coincides with the period when air temperatures are also highest, and therefore ablation is highest, so that mass balance relationships are complex on these glaciers.

The glaciers of the Cordillera Blanca in Peru provide a good example of tropical glaciation (Figure 2.9). Here there is more than 720 km^2 of high-altitude glaciers, making it the largest glaciated area in the tropics, equivalent to ~25% of the area of tropical glaciers globally. Glaciers exist in the mountains for 120 km between 8.5° and 10°S along a northwest–southeast strike fault, dividing stream runoff between the Pacific and Atlantic oceans. Between 70 and 90% of the annual precipitation falls between October and March in these regions. The glaciers are important socially and economically because they act as water stores and therefore influence water supplies. River water for drinking and sanitation comes exclusively from glacier melt during the dry season when there is little or no precipitation. The rivers are also increasingly used for hydroelectric power generation.

Glaciers exist here because the air temperatures are so cold at high elevations (up to 6000 m above sea level). In contrast to mid-latitude glaciers, tropical glaciers do not have summer time melt seasons characterised by widespread above-freezing air temperature. Their lower altitude portions are warmed directly by year-round exposure to above-freezing air but at higher altitudes absorption of sunlight ultimately supplies all the energy that sustains ablation. Ablation is sensitive to the amount of absorbed solar radiation, air temperature, atmospheric humidity, cloud cover and wind. Tropical glaciers therefore react faster to changes in ablation than those in the mid-latitudes, such that a number of glaciers have receded dramatically or even disappeared in recent years. The tropical glaciers of the Cordillera Blanca have been receding rapidly in recent years, with a consequent reduction in glacier volume throughout the nineteenth and twentieth centuries. Larger-scale climate features, such as the El Niño-Southern Oscillation (ENSO) have been shown to have an impact on interannual variability of temperature, precipitation and stream discharge in this region. The mostly widely cited explanation for this recession is a reduction in air humidity, with all the consequent changes in energy and mass balance. Rising air temperatures explain only part of the glacier recession.

Figure 2.9   Mountain glacier in the Cordillera Blanca, Peru. Note the progressive change from clean ice in the steep icefall to debris cover near the snout where there is a much lower surface gradient. [Photograph: N.F. Glasser.]

The sediments and landforms created by tropical glaciers are varied. Most glaciers in the tropical Andes are receding and undergoing down-wasting in their terminal parts. As this debris becomes concentrated on the ice surface, the glaciers become progressively more debris-mantled down-glacier. The debris forms an irregular cover of hummocky, coarse angular material interspersed with supraglacial ponds, much like those developed on Himalayan glaciers (see Section 2.5). In front of some glaciers are very large (up to 100 m high) lateral and frontal moraines, dating from the Little Ice Age or earlier (Figure 2.10). The moraines often have vegetated outer faces and an unvegetated, loose, collapsing inner face composed of sandy boulder-gravel derived from a mixture of rockfall debris and material from the zone of traction along the valley sides. Like their Himalayan counterparts, these terminal moraines often impound large supraglacial or proglacial lakes that can drain catastrophically if the moraine dam fails. The moraines that dam these proglacial lakes therefore pose an increasing hazard to communities in the Andes.

Figure 2.10 Cirque glacier in the Cordillera Blanca, Peru. The glacier has receded behind the large moraines in the foreground during the twentieth and twenty-first centuries. The moraines are now being dissected by the proglacial stream fed by the glacier. [Photograph: N.F. Glasser]

The moraines are prone to failure through collapse, overtopping by lake waters or the effect of displacement waves resulting from ice and rock avalanches.

2.7 ARCTIC POLYTHERMAL GLACIERS

Glaciers are common on many of the landmasses surrounding the Arctic Ocean. Excluding the Greenland Ice Sheet, glaciers and ice caps cover about 275 000 km^2 of the archipelagos of the Canadian, Norwegian and Russian High Arctic and the area north of about 60°N in Alaska, Iceland and Scandinavia. Although there is of course huge variability in their behaviour it appears that since the 1940s most Arctic glaciers have experienced a predominantly negative net surface mass balance. Indeed, there is evidence in the form of Arctic *ice shelf break-up* that recent environmental change is accelerating. Until recently, a large ice shelf (the Ward Hunt Ice Shelf: 443 km^2 in area) existed along the northern coast of Ellesmere Island in Canada's Nunavut territory. After a 30 year decline in extent, the ice shelf suddenly broke-up between 2000 and 2002. The ice shelf fragmented into two main parts with many additional fissures and calved a number of ice islands. This marked the disappearance of the largest ice shelf in the Arctic.

One of the most intensely studied areas of the Arctic is the Svalbard archipelago (77° to 80°N), at the northern extent of the mild Norwegian Current, a branch of the Gulf Stream. The archipelago enjoys a relatively mild climate for its northern latitude.

On the west coast, the average annual temperature is –6°C. The average temperature on the west coast in the warmest month (July) is 5°C, whereas in the coldest month (January) it is –15°C. Although there are contrasts between the maritime west coast and the more continental interior, precipitation in Svalbard is generally low. Typical values at sea level are 400–600 mm annually, falling to half these values inland. Precipitation in the more mountainous regions is increased by orographic effects, but even on the glaciers snowfall of more than 2–4 m is rare. Ice-free land areas are underlain by permafrost to depths of between 100 and 400 m. In this sense Svalbard is typical of the Arctic, in that precipitation is generally low and declines linearly with elevation.

There are of course local variations in glacier type and morphology in the Arctic, but it is possible to make the following generalisations based on Svalbard glaciers.

1. Many Arctic glaciers are *polythermal*; that is the snout, lateral margins and surface layer of the glacier are below the pressure-melting point, whereas thicker, higher-level ice in the accumulation area is often warm-based (see Section 3.4). This mix of thermal regimes makes the dynamics of polythermal glaciers complex. The typical polythermal glacier moves by sliding on its bed or by subsole deformation where warm-based ice dominates in the accumulation zone, and moves only by internal deformation where it is cold-based ice at the snout and lateral margins. Meltwater tends to follow supra- and englacial routes and well-developed basal hydrological networks are rare. As a result, polythermal glaciers tend to carry a high basal debris load, with debris-rich basal ice zones between 1 and 3 m thick, and debris concentrations of up to 50%. Their surfaces rarely have a substantial cover of debris, although medial moraines are commonly observed in their lower ablation areas (Figure 2.11).

2. Svalbard is famous for its high proportion of *surge-type glaciers* (Box 2.4). An estimated 35% of the glaciers on Svalbard are surge-type. These glaciers are prone to dramatic increases in velocity and rapid frontal advances, followed by periods of quiescence during which velocities are generally low. Surge-type glaciers in Svalbard typically have relatively long quiescent phases (10–200 years) between short-lived surge events (1–5 years).

3. *Superimposed ice* formed by the refreezing of melted snow is common on the surface of these glaciers. This commonly forms on the glacier surface if there are short periods of positive air temperatures in early winter, often coinciding with rainfall, which cause rapid glacier-wide melting. Percolating water then refreezes to form superimposed ice on the lower half of the glacier, and wetted-refrozen snow and ice lenses at higher altitudes. The formation of super-imposed ice has significant implications for glacier mass balance, because it locally comprises up to 20% of winter balances and accounts for between 16 and 25% of the annual accumulation.

4. There is evidence that the sediments and landforms produced by Svalbard glaciers are strongly influenced by *structural glaciological controls* on debris entrainment and transport (Figure 2.12). *Moraine-mound complexes* in front of receding Svalbard glaciers are formed by proglacial and englacial thrusting, folding and deformation of the bed during surges. On a smaller scale, a range of glacier structures arrange debris into foliation-parallel ridges, supraglacial debris stripes and geometrical ridge networks (see Section 7.5).

Figure 2.11   Aerial photograph of the glacier Kongsvegen in Svalbard. [Image courtesy of: Norsk Polarinstittut]

BOX 2.4: SURGE-TYPE GLACIERS IN SVALBARD

An estimated 35% of the glaciers on Svalbard are surge-type; that is they are prone to dramatic increases in velocity and rapid frontal advances, followed by periods of quiescence during which velocities are generally low. Surge-type glaciers in Svalbard typically have relatively long quiescent phases (10–200 years) between short-lived surge events (1–5 years). Jiskoot *et al.* (2000) analysed statistically the possible controls on the distribution of surge-type glaciers in Svalbard using a large number of glacial and geological attributes. They looked at the potential effects of geology, mass-balance conditions and thermal regime on surging. They concluded from their analysis that long polythermal glaciers with relatively steep slopes overlying young fine-grained sedimentary lithologies are most likely to be of surge type in Svalbard. They found no statistical relationship between geology and surge-type behaviour. Possible explanations for the importance of glacier length are transport-distance-related substrate properties, distance-related attenuation of longitudinal stresses and the relationship between glacier size and polythermal glacier regime. The overall conclusion of their study is that the surge-type potential of Svalbard glaciers is greatest for polythermal glaciers overlying fine-grained potentially deformable beds. The photograph over the page shows the snout of a small surging glacier (Arebreen) in Svalbard.

Source: Jiskoot, H., Murray, T. and Boyle, P. (2000) Controls on the distribution of surge-type glaciers in Svalbard. *Journal of Glaciology*, **46**, 412–22. [Photograph: M.R. Bennett]

Figure 2.12   Hummocky moraines in front of a receding glacier in Svalbard, interpreted as the product of englacial thrusting within the ice. [Photograph: N.F. Glasser]

5. Deformation of permafrost is also important in the formation of ice-cored moraines and push moraines (see Section 9.1.1). Stresses beneath the advancing glaciers are transmitted to the proglacial sediments and can be sufficient to cause proglacial deformation of the permafrost layer. Folding, thrust-faulting and overriding of proglacial sediments are also possible under these conditions. The nature of the deformation is controlled by the mechanical properties of the sediment, which is influenced by the water content and thermal condition, whether it is frozen or unfrozen.

2.8 SUMMARY

Glacier ice is at present widely distributed across the globe. Glaciers can form at any location where accumulation exceeds ablation on a multi-annual basis. As a result, there is a huge variety in the types of glaciers that are found on Earth today. Large ice sheets and glaciers exist close to sea level in both polar regions, in the mid-latitudes where there is sufficient precipitation and even in tropical regions at high altitude. There is also a huge variety in the landforms and sediments produced by these glaciers. Understanding the behaviour of contemporary ice sheets and glaciers, as well as the origins of the landforms and sediments that they produce, is important because these are the analogues that we use to interpret the products of former ice sheets and glaciers.

SUGGESTED READING

Barry (2006) reviews the current response of the world's glaciers to recent climate change.

Antarctica: The results of Antarctic ice-core drilling are presented by the EPICA (European Project for Ice Coring in Antarctica) Community Members (2004). Bamber *et al.* (2000) outline the large-scale flow of the Antarctic Ice Sheet. Siegert (2000) reviews evidence for Antarctic subglacial lakes, and Fricker *et al.* (2007) describe subglacial lake connections. Bennett (2003) provides a review of ice stream behaviour, and Ng and Conway (2004) provide a good example of recent ice stream stagnation. Oppenheimer (1998) reviews some of the uncertainties in assessing the future behaviour of the ice sheet. Scambos *et al.* (2000) identify the links between climate warming and the break-up of ice shelves on the Antarctic Peninsula. Pritchard and Vaughan (2007) describe the effects of ice-shelf removal on glacier velocities. Shepherd and Wingham (2007) consider the sea-level contribution of the ice sheet. Fitzsimons (1997, 2003), Fitzsimons *et al.* (2008) and Atkins *et al.* (2002) have described erosional and depositional processes and landforms associated with the cold-based glaciers of the Dry Valleys

(Antartica). Johnson *et al.* (2008) illustrate how cosmogenic isotope dating can be used to infer past Antarctic ice thickness.

Greenland: The changes in the behaviour of the outlet glaciers that drain the Greenland Ice Sheet are described in a series of papers by Joughin *et al.* (2004, 2008a,b), Howat *et al.* (2005, 2007), Luckman and Murray (2005), Luckman *et al.* (2006), Moon and Joughin (2008) and Rignot and Kanagaratnam (2006). Krabill *et al.* (2004) document peripheral thinning of the ice sheet. The role of surface meltwater in this speed-up is highlighted by Zwally *et al.* (2002). Das *et al.* (2008) document the role of supraglacial lake drainage through crevasses in providing meltwater to the bed. Changes due to tidewater glacier behaviour are discussed by Howat *et al.* (2005) and Pfeffer (2007).

Patagonia and New Zealand: Rivera *et al.* (2007) explore the relationship between climate and glacier response in Patagonia. Hochstein *et al.* (1995), Kirkbride and Warren (1999) and Anderson *et al.* (2006) explore similar relationships for New Zealand glaciers. Fitzharris *et al.* (2007) have investigated climatic teleconnections between glaciers in both areas. Davies *et al.* (2003) describe the role of outburst floods in transferring sediments to the surface of the Franz Josef Glacier in New Zealand, and Hambrey and Ehrmann (2004) describe the broader-scale depositional products around a number of New Zealand glaciers. In Patagonia, Glasser and Hambrey (2002) outline the landforms and sediments developed at the margins of one of the temperate outlet glaciers from the North Patagonian Icefield. For comparison, Goodsell *et al.* (2005) have described how debris is entrained and transported by a European alpine glacier.

Alaska and Iceland: Surge-type glaciers in Alaska are described in a number of papers; of particular note, however, is the work of Kamb *et al.* (1985), Raymond (1987), Humphrey and Raymond (1994) and Eisen *et al.* (2001). Truffer *et al.* (2000, 2001) describe studies of Black Rapids Glacier in Alaska using a range of invasive monitoring techniques. Papers relevant to Matanuska Glacier include those by Lawson (1982) on flow tills and Lawson *et al.* (1998) on the the effects of glaciohydraulic supercooling. For the wider context of the controversy surrounding supercooling see Alley *et al.* (1998) and Cook *et al.* (2006). The landform assemblages associated with both surge and non-surge type Icelandic glaciers are well document in the volume edited by Evans (2003). Evans and Twigg (2002) document in detail the landform assemblages associated with these glaciers.

Himalayas: Luckman *et al.* (2007) provide data on the measured flow rates of glaciers in the Everest region. Benn and Owen (2002), Benn *et al.* (2003), Owen *et al.* (2003) and Hambrey *et al.* (2009) have all described various aspects of the glacial sedimentary environments in different areas of the Himalaya. Richardson and Reynolds (2000) provide a review of glacial hazards in the Himalaya, and Quincey *et al.* (2007) describe how remote-sensing data sets can be used to monitor the development of glacial lake hazards in the terminal areas of these glaciers. Benn *et al.* (2001) document the evolution of supraglacial lakes on the

Ngozumpa Glacier in Nepal, and Gully and Benn (2007) explored the englacial drainage systems of Himalayan glaciers. Cenderelli and Wohl (2001) described the geomorphological effects of two former outburst floods in the Mount Everest region of Nepal.

Cordillera Blanca, Peru: The book by Kaser and Osmaston (2002) provides excellent coverage of all aspects of tropical glaciers. The relationship between climate and glaciation at low latitudes is covered by Kaser (2001). Francou *et al.* (2003) review the glacier–climate interactions for South American tropical glaciers, and Kaser *et al.* (2004) and Cullen *et al.* (2006) do the same for glaciers on Kilimanjaro in Africa. Ribstein *et al.* (2005) and Mark and Seltzer (2003) provide studies of the contemporary relationships between climate, glacier mass balance and glacier hydrology in the Andes of South America. Kaser (1999) reviews current knowledge of the recent fluctuations of tropical glaciers. Hubbard *et al.* (2005) describe the hazards imposed by moraine-dammed lakes in the tropical Andes.

Arctic Polythermal Glaciers: Dowdeswell *et al.* (1997) review the relationship between Arctic glaciers and recent climate change, and Mueller *et al.* (2003) describe the break-up of the largest ice shelf in the Arctic. Copland and Sharp (2001) and Copland *et al.* (2003) explain subglacial and hydrological conditions beneath a polythermal glacier on Ellesmere Island. Hamilton and Dowdeswell (1996) outline the controls on surge-type glaciers in Svalbard. Wadham and Nuttall (2002) and Wright *et al.* (2007) studied the formation of superimposed ice and its significance for Svalbard glaciers. General reviews of the sediments and landforms at polythermal glaciers are provided by Glasser and Hambrey (2003) for Svalbard and by Ó Cofaigh *et al.* (2003) for the Canadian and Greenland High Arctic. The influence of structural glaciology on sediment–landform associations on Svalbard is outlined by Glasser *et al.* (1998), Hambrey *et al.* (1999) and Glasser and Hambrey (2001). Etzelmüller *et al.* (1996) outline glacier–permafrost interactions and Boulton *et al.* (1996, 1999) describe sediments and landforms in front of receding glaciers in Svalbard.

Alley, R.B., Lawson, D.E., Evenson, E.B., *et al.* (1998) Glaciohydraulic supercooling: a freeze-on mechanism to create stratified, debri-rich basal ice. II. Theory. *Journal of Glaciology*, **44**, 563–9.

Anderson, B., Lawson, W., Owens, I. and Goodsell, B. (2006) Past and future mass balance of 'Ka Roimata o Hine Hukatere' Franz Josef Glacier, New Zealand. *Journal of Glaciology*, **52**, 597–607.

Atkins, C.B., Barrett, P.J. and Hicock, S.R. (2002) Cold glaciers erode and deposit: evidence from Allan Hills, Antarctica. *Geology*, **30**, 659–62.

Bamber, J.L., Vaughan, D.G. and Joughin, I. (2000) Widespread complex flow in the interior of the Antarctic Ice Sheet. *Science*, **287**, 1248–50.

Barry, R.G. (2006) The status of research on glaciers and global glacier recession: a review. *Progress in Physical Geography*, **30**, 285–306.

Benn, D.I. and Owen, L.A. (2002) Himalayan glacial sedimentary environments: a framework for reconstructing and dating the former extent of glaciers in high mountains. *Quaternary International*, **97–98**, 3–25.

Benn, D.I., Wiseman, S. and Hands, K.A. (2001) Growth and drainage of supraglacial lakes on the debris mantled Ngozumpa Glacier, Khumbu Himal, Nepal. *Journal of Glaciology*, **47**, 626–38.

Benn, D.I., Kirkbride, M.P., Owen, L.A. and Brazier, V. (2003) Glaciated valley landsystems, in Glacial Landsystems (ed. D.J.A. Evans), Arnold, London, pp. 372–406.

Bennett, M.R. (2003) Ice streams as the arteries of an ice sheet: their mechanics, stability and significance. *Earth Science Reviews*, **61**, 309–39.

Boulton, G.S., Van der Meer, J.J.M., Hart, J., *et al.* (1996) Till and moraine emplacement in a deforming bed surge - an example from a marine environment. *Quaternary Science Reviews*, **15**, 961–87.

Boulton, G.S., Van der Meer, J.J.M., Beets, D.J., *et al.* (1999) The sedimentary and structural evolution of a recent push moraine complex: Holmstrømbreen, Spitsbergen. *Quaternary Science Reviews*, **18**, 339–71.

Cenderelli, D.A. and Wohl, E.E. (2001) Flow hydraulics and geomorphic effects of glacial-lake outburst floods in the Mount Everest region, Nepal. *Earth Surface Processes and Landforms*, **28**, 385–407.

Cook, S.J., Waller, R.I. and Knight, P.G. (2006) Glaciohydraulic supercooling: the process and its significance. *Progress in Physical Geography*, **30**, 577–88.

Copland, L. and Sharp, M. (2001) Mapping thermal and hydrological conditions beneath a polythermal glacier with radio-echo sounding. *Journal of Glaciology*, **47**, 232–42.

Copland, L., Sharp, M.J. and Nienow, P.W. (2003) Links between short-term velocity variations and the subglacial hydrology of a predominantly cold polythermal glacier. *Journal of Glaciology*, **49**, 337–48.

Cullen, N.J., Mölg, T., Kaser, G., *et al.* (2006) Kilimanjaro Glaciers: recent areal extent from satellite data and new interpretation of observed 20th century retreat rates. *Geophysical Research Letters*, **33**, L16502.

Das, S.B., Joughin, I., Behn, M.D., *et al.* (2008) Fracture propagation to the base of the Greenland ice sheet during supraglacial lake drainage. *Science*, **320**, 778–81.

Davies, T.R.H., Smart, C.C. and Turnbull, J.M. (2003) Water and sediment outbursts from advanced Franz Josef Glacier, New Zealand. *Earth Surface Processes and Landforms*, **28**, 1081–96.

Dowdeswell, J.A., Hagen, J.O., Björnsson, H., *et al.* (1997) The mass balance of Circum-Arctic Glaciers and recent climate change. *Quaternary Research*, **48**, 1–14.

Eisen, O., Harrison, W.D. and Raymond, C.F. (2001) The surges of Variegated Glacier, Alaska, USA, and their connection to climate and mass balance. *Journal of Glaciology*, **47**, 351–8.

EPICA Community Members (2004) Eight glacial cycles from an Antarctic ice core. *Nature*, **429**, 623–8.

Etzelmüller, B., Hagen, J.O., Vatne, G., *et al.* (1996) Glacial debris accumulation and sediment deformation influenced by permafrost: examples from Svalbard. *Annals of Glaciology*, **22**, 53–62.

Evans, D.J.A. (ed.) (2003) *Glacial Landsystems*, Arnold, London.

Evans, D.J.A. and Twigg, D.R. (2002) The active temperate glacial landsystem: a model based on Breiðamerkurkull and Fjallsjökull, Iceland. *Quaternary Science Reviews*, **21**, 2143–77.

Fitzharris, B.B., Clare, G.R. and Renwick, J. (2007) Teleconnections between Andean and New Zealand glaciers. *Global and Planetary Change*, **59**, 159–74.

Fitzsimons, S.J. (1997) Depositional models for moraine formation in East Antarctic coastal oases. *Journal of Glaciology*, **43**, 256–64.

Fitzsimons, S.J. (2003) Ice-marginal terrestrial landsystems: polar-continental glacier margins, in *Glacial Landsystems* (ed. D.J.A. Evans), Arnold, London, pp. 89–110.

Fitzsimons, S., Webb, N., Mager, S. *et al.* (2008) Mechanisms of basal ice formation in polar glaciers: an evaluation of the apron entrainment model. *Journal of Geophysical Research*, **113** (F02010), doi:10.1029/2006JF000698.

Francou, B., Vuille, M., Wagnon, P., *et al.* (2003) Tropical climate change recorded by a glacier in the central Andes during the last decades of the twentieth century: Chacaltaya, Bolivia, 16°S. *Journal of Geophysical Research*, **108** (D5), 4154.

Fricker, H.A., Scambos, T.A., Bindschadler, R. and Padman, L. (2007) An active subglacial water system in West Antarctica mapped from space. *Science*, **315**, 1544–8.

Glasser, N.F. and Hambrey, M.J. (2001) Styles of sedimentation beneath Svalbard valley glaciers under changing dynamic and thermal regimes. *Journal of the Geological Society, London*, **158**, 697–707.

Glasser, N.F. and Hambrey, M.J. (2002) Sedimentary facies and landform genesis at a temperate outlet glacier: Soler Glacier, North Patagonian Icefield. *Sedimentology*, **49**, 43–64.

Glasser, N.F. and Hambrey, M.J. (2003) Ice-marginal terrestrial landsystems: Svalbard polythermal glaciers, in *Glacial Landsystems* (ed. D.J.A. Evans), Arnold, London, pp. 65–88.

Glasser, N.F., Hambrey, M.J., Crawford, K.R., *et al.* (1998) The structural glaciology of Kongsvegen, Svalbard, and its role in landform genesis. *Journal of Glaciology*, **44**, 136–48.

Goodsell, R.C., Hambrey, M.J. and Glasser, N.F. (2005) Debris transport in a temperate valley glacier: Haut Glacier d'Arolla, Valais, Switzerland. *Journal of Glaciology*, **51**, 139–46.

Gulley, J. and Benn, D.I. (2007) Structural control of englacial drainage systems in Himalayan debris-covered glaciers. *Journal of Glaciology*, **53**, 399–412.

Hambrey, M.J. and Ehrmann, W. (2004) Clast-shape and textural characteristics of glacigenic sediments in high-alpine glacierised catchments: Mount Cook area, New Zealand. *Boreas*, **33**, 300–18.

Hambrey, M.J., Bennett, M.R., Dowdeswell, J.A., *et al.* (1999) Debris entrainment and transfer in polythermal glaciers. *Journal of Glaciology*, **45**, 69–86.

Hambrey, M.J., Quincey, D.J., Glasser, N.F., *et al.* (2009). Sedimentological, geomorphological and dynamic context of debris-mantled glaciers, Mount Everest (Sagarmartha) region, Nepal. *Quaternary Science Reviews*, **27**(25–26), 2361–89.

Hamilton, G.S. and Dowdeswell, J.A. (1996) Controls on glacier surging in Svalbard. *Journal of Glaciology*, **42**, 157–68.

Hochstein, M.P., Claridge, D., Henrys, S.A., *et al.* (1995). Downwasting of the Tasman Glacier, South Island, New Zealand: changes in the terminus region between 1971 and 1993. *New Zealand Journal of Geology and Geophysics*, **38**, 1–16.

Howat, I.M., Joughin, I., Tulaczyk, S. and Gogineni, S. (2005) Rapid retreat and acceleration of Helheim Glacier, East Greenland. *Geophysical Research Letters*, **32** (L22052).

Howat, I.M., Joughin, I. and Scambos, T.A. (2007) Rapid changes in ice discharge from Greenland Outlet Glaciers. *Science*, **315**, 1559–61.

Hubbard, B., Heald, A., Reynolds, J.M., *et al.* (2005) Impact of a rock avalanche on a moraine-dammed proglacial lake: Laguna Safuna Alta, Cordillera Blanca, Peru. *Earth Surface Processes and Landforms*, **30**, 1251–64.

Humphrey, N.F. and Raymond, C.F. (1994) Hydrology, erosion and sediment production in a surging glacier – Variegated Glacier, Alaska, 1982–83. *Journal of Glaciology*, **40**, 539–52.

Johnson, J.S., Bentley, M.J. and Gohl, K. (2008) First exposure ages from the Amundsen Sea Embayment, West Antarctica: the late Quaternary context for recent thinning of Pine Island, Smith and Pope Glaciers. *Geology*, **36**, 223–6.

Joughin, I., Abdalati, W. and Fahnestock, M. (2004) Large fluctuations in speed on Greenland's Jakobshavn Isbræ glacier. *Nature*, **432**, 608–10.

Joughin, I., Das, S.B., King, M.A., *et al.* (2008a) Seasonal speedup along the western flank of the Greenland Ice Sheet. *Science*, **5877**, 781–3.

Joughin, I., Howat, I., Alley, R.B., *et al.* (2008b) Ice-front variation and tidewater behaviour on Helheim And Kangerdlugssuaq Glaciers, Greenland. *Journal of Geophysical Research*, **113** (F01004), doi:10.1029/2007JF000837.

Kamb, B., Raymond, C.F., Harrison, W.D., *et al.* (1985) Glacier surges mechanism: 1982–1983 surge of Variegated Glacier, Alaska. *Science*, **227**, 469–79.

Kaser, G. (1999) A review of the modern fluctuations of tropical glaciers. *Global and Planetary Change*, **22**, 93–103.

Kaser, G. (2001) Glacier–climate interactions at low latitudes. *Journal of Glaciology*, **47**, 195–204.

Kaser, G. and Osmaston, H. (2002) *Tropical Glaciers*, Cambridge University Press, Cambridge.

Kaser, G., Hardy, D.R., Mölg, T., *et al.* (2004) Modern glacier retreat on Kilimanjaro as evidence of climate change: observations and facts. *International Journal of Climatology*, **24**, 329–39.

Kirkbride, M.P. and Warren, C.R. (1999) Tasman Glacier, New Zealand: 20th-century thinning and predicted calving retreat. *Global and Planetary Change*, **22**, 11–28.

Krabill, W., Hanna, E., Huybrechts, P. *et al.* (2004) Greenland Ice Sheet: increased coastal thinning. *Geophysical Research Letters*, **31,** (L24402).

Lawson, D.E. (1982) Mobilisation, movement and deposition of active subaerial sediment flows, Matanuska glacier, Alaska. *Journal of Geology*, **90**, 279–300.

Lawson, D.E., Strasser, J.C., Evenson, E.B., *et al.* (1998) Glaciohydraulic supercooling: a freeze-on mechanism to create stratified, debris-rich basal ice.I. Field Evidence. *Journal of Glaciology*, **44**, 547–62.

Luckman, A. and Murray, T. (2005) Seasonal variation in velocity before retreat of Jakobshavn Isbræ, Greenland. *Geophysical Research Letters*, **32**, (L08501).

Luckman, A., Murray, T., de Lange, R. and Hanna, E. (2006) Rapid and synchronous ice-dynamic changes in East Greenland. *Geophysical Research Letters*, **33,** (L03503).

Luckman, A., Quincey, D. and Bevan, S. (2007) The potential of satellite radar interferometry and feature tracking for monitoring flow rates of Himalayan glaciers. *Remote Sensing of Environment*, **111**, 172–81.

Mark, B.G. and Seltzer, G.O. (2003) Tropical glacier meltwater contribution to stream discharge: a case study in the Cordillera Blanca, Peru. *Journal of Glaciology*, **49**, 271–81.

Moon, T. and Joughin, I. (2008) Changes in ice front position on Greenland's outlet glaciers from 1992 to 2007. *Journal of Geophysical Research*, **113** (F02022), doi:10.1029/2007JF000927.

Mueller, D.R., Vincent, W.F. and Jeffries, M.O. (2003) Breakup of the largest Arctic ice shelf and associated loss of an epishelf lake. *Geophysical Research Letters*, **30**, doi:10.1029/2003GL017931.

Ng, F. and Conway, H. (2004) Fast-flow signature in the stagnated Kamb Ice Stream, West Antarctica. *Geology*, **32**, 481–4.

Ó Cofaigh, C., Evans, D.J.A. and England, J. (2003) Ice-marginal terrestrial landsystems: Sub-polar glacier margins of the Canadian and Greenland High Arctic, in *Glacial Landsystems* (ed. D.J.A. Evans), Arnold, London, pp. 44–64.

Oppenheimer, M. (1998) Global warming and the stability of the West Antarctic Ice Sheet. *Nature*, **393**, 325–32.

Owen, L.A., Derbyshire, E. and Scott, C.H. (2003) Contemporary sediment production and transfer in high-altitude glaciers. *Sedimentary Geology*, **155**, 13–36.

Pfeffer, W.T. (2007) A simple mechanism for irreversible tidewater glacier retreat. *Journal of Geophysical Research*, **112** (F03S25), doi:10.1029/2006JF000590.

Pritchard, H.D. and Vaughan, D.G. (2007) Widespread acceleration of tidewater glaciers on the Antarctic Peninsula. *Journal of Geophysical Research*, **112** (F03S29), doi:10.1029/2006JF000597.

Quincey, D.J., Richardson, S.D., Luckman A., *et al.* (2007) Early recognition of glacial lake hazards in the Himalaya using remote sensing datasets. *Global and Planetary Change*, **56**, 137–52.

Raymond, C.F. (1987) How do glaciers surge? A review. *Journal of Geophysical Research*, **92**, 9121–34.

Ribstein, P., Tiriau, E., Francou, B. and Saravia, R. (1995) Tropical climate and glacier hydrology: a case study in Bolivia. *Journal of Hydrology*, **165**, 221–34.

Richardson, S.D. and Reynolds, J.M. (2000) An overview of glacial hazards in the Himalayas. *Quaternary International*, **65/66**, 31–47.

Rignot, E. and Kanagaratnam, P. (2006) Changes in the velocity structure of the Greenland Ice Sheet. *Science*, **311**, 986–90.

Rivera, A., Benham, T., Casassa, G., *et al.* (2007) Ice elevation and areal changes of glaciers from the northern Patagonia icefield, Chile. *Global and Planetary Change*, **59**, 126–37.

Scambos, T.A., Hulbe, C., Fahnestock, M. and Bohlander, J. (2000) The link between climate warming and break-up of ice shelves in the Antarctic peninsula. *Journal of Glaciology*, **46**, 516–30.

Shepherd, A. and Wingham, D. (2007) Recent sea-level contributions of the Antarctic and Greenland Ice Sheets. *Science*, **315**, 1529–32.

Siegert, M.J. (2000) Antarctic subglacial lakes. *Earth Science Reviews*, **50**, 29–50.

Truffer, M., Harrison, W.D. and Echelmeyer, K.A. (2000) Glacier motion dominated by processes deep in underlying till. *Journal of Glaciology*, **46**, 213–21.

Truffer, M., Echelmeyer, K.A. and Harrison, W.D. (2001) Implications of till deformation on glacier dynamics. *Journal of Glaciology*, **47**, 123–34.

Wadham, J.L. and Nuttall, A.-M. (2002) Multiphase formation of superimposed ice during a mass-balance year at a maritime high-Arctic glacier. *Journal of Glaciology*, **48**, 545–51.

Wright, A.P., Wadham, J.L., Siegert, M.J., *et al*. (2007) Modeling the refreezing of meltwater as superimposed ice on a high Arctic glacier: a comparison of approaches. *Journal of Geophysical Research*, **112** (F04016), doi:10.1029/2007JF000818.

Zwally, H.J., Abdalati, W., Herring, T., *et al*. (2002) Surface melt-induced acceleration of Greenland ice-sheet flow. *Science*, **297**, 218–22.

3: Mass Balance and the Mechanisms of Ice Flow

In this chapter we look at glacier mass balance and the mechanisms of ice flow and consider the formation of glaciers and glacier ice, as well as how glaciers flow and the temperature distribution within them. We also investigate why some glaciers flow faster and are more active than others.

3.1 ANNUAL MASS BALANCE

A glacier will form whenever a body of snow accumulates, compacts and turns to ice. This can occur in any climatic zone where the input of snow exceeds the rate at which it melts. The length of time required to form a glacier will depend on the rate at which the snow accumulates and turns to ice. If this rate of *accumulation* is high and the loss due to melting is low then a glacier will form quickly. Once established its survival will depend on the balance between accumulation and melting (*ablation*). This balance between accumulation and ablation is known as the *mass balance* of the glacier and is largely dependent on climate. Understanding mass balance is important because this determines the net gain or loss of ice on glaciers and ice sheets, which has important implications for global sea-level change (Box 3.1).

BOX 3.1: THE MASS BALANCE OF THE POLAR ICE SHEETS AND GLOBAL SEA-LEVEL RISE

One of the most important advances in glaciology in the past two decades has been the ability to monitor how the Earth's two polar ice sheets are changing using satellite measurements. As global temperatures have risen, so have rates of snow accumulation, ice melting and glacier flow in the Polar Regions. Although

---

the balance between these opposing processes has varied considerably on a regional scale, satellite data show that the large Antarctic and Greenland Ice Sheets are each losing mass overall. Rignot and Thomas (2002) used satellite data to determine the mass balance of the two ice sheets and showed that: (i) the Greenland Ice Sheet is losing mass by near-coastal thinning; and (ii) the West Antarctic Ice Sheet, with thickening in the west and thinning in the north, is thinning overall. Shepherd and Wingham (2007) later estimated that the combined imbalance of the two ice sheets is about 125 gigatons (1.25×10^{11} tons) per year of ice, enough to raise sea level by 0.35 mm per year. This is only about 10% of the observed present rate of sea-level rise of 3.0 mm per year. This means that the Earth's smaller mountain glaciers and ice caps, which react more quickly to changes in mass balance, must be contributing the largest amounts of water to the oceans (Meier *et al.*, 2007). However, much of the loss from Antarctica and Greenland is the result of the flow of ice to the ocean from ice streams and outlet glaciers. As this has accelerated over the past decade, we might expect to see a rise in the sea-level contribution from the two polar ice sheets over the course of the twenty-first century (see Sections 2.1 and 2.2). In both polar continents there are suspected triggers for the accelerated ice discharge, namely surface and ocean warming.

Sources: Meier, M.F., Dyurgerov, M.B., Rick, U.K., *et al.* (2007) Glaciers dominate eustatic sea-level rise in the 21st Century. *Science*, **317**, 1064–7. Rignot, E. and Thomas, R.H. (2002) Mass balance of polar ice sheets. *Science*, **297**, 1502–6. Shepherd, A. and Wingham, D. (2007) Recent sea-level contributions of the Antarctic and Greenland Ice Sheets. *Science*, **315**, 1529–32.

Figure 3.1 shows how the total amount of accumulation and the total amount of ablation each year defines the mass balance of a glacier. Ablation will tend to dominate in the warm summer months and accumulation in the winter months. If the amount of ablation equals the amount of accumulation over a year the *net balance* of the glacier will be zero and its size will remain constant (Figure 3.1). On the other hand, if there is more accumulation than ablation then the net balance will be positive and the glacier will grow and expand. If it is has a negative mass balance then the glacier will gradually disappear.

The study of *glacier mass balance* is, therefore, the study of inputs and outputs to the glacial ice system. Inputs to a glacier's mass balance include: snow, hail, frost, avalanched snow and rainfall. If these inputs survive summer ablation they will begin a process of transformation into glacier ice. The term *firn* or *névé* is used for snow that has survived a summer melt season and has begun this transformation. The transformation involves: (i) compaction; (ii) the expulsion of air; and (iii) the growth of an interlocking system of ice crystals. Dry fresh snow is about 97% air by volume and has a density of 100 kg m^{-3} while glacier ice has almost no air within it and a density of 900 kg m^{-3}. The rate at which this transformation takes place is dependent on climate. If snow fall is high and significant melting occurs then the

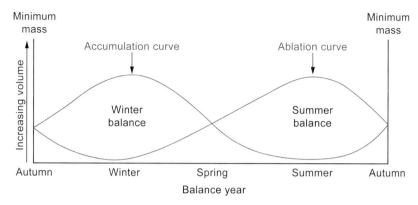

Figure 3.1 Accumulation and ablation curves define the mass balance year for a glacier. The winter balance is positive and the summer is negative. If the winter and summer balances are exactly equal, then the net mass balance will be zero and the glacier will neither advance or retreat. [Modified from: Sugden and John (1976) *Glaciers and Landscape*, Edward Arnold, figure 3.2, p. 37]

process can be rapid, because older snow is quickly buried by fresh accumulation, which compacts the firn, while alternate melting and refreezing encourages the growth of new ice crystals. In contrast, where accumulation rates are low and little melting occurs the transformation can be extremely slow. For example, in the interior of the East Antarctic Ice Sheet, where there is very little accumulation and even less melt, the transformation may take up to 3500 years. By contrast on the Seward glacier in Alaska the transformation is achieved in as little as 3–5 years, due to the high accumulation and melt rates.

Outputs from the mass balance system are collectively termed *ablation*. This ablation can occur in three ways, by: ice melt, iceberg calving and sublimation. *Glacial meltwater* is derived from direct melting of ice on the surface of, or within, a glacier. On the surface this is a function of solar radiation received, whereas within and at the base of the glacier heat is supplied by: (i) friction due to ice flow; and (ii) by heat derived from the Earth's crust beneath the glacier (*geothermal heat*). Melting therefore can occur both on the surface of the glacier and within the glacier itself. Surface melting is primarily a result of warm air temperatures and is therefore highly seasonal, whereas melting within the glacier is not. It is important to emphasise that melting is not simply confined to the ice margin but may occur across the whole of the glacier surface.

Where glaciers terminate in water, either in the sea or in a lake, blocks of ice will break from the front (*snout* or *terminus*) of the glacier as icebergs (Figure 3.2). This process is known as *iceberg calving*. It can be a particularly rapid way of losing mass from a glacier. In very cold and dry environments mass may also be lost through *sublimation*, which is the direct evaporation of water from its solid state as ice.

In summary, it is the relative balance between inputs and outputs to a glacier that determine its mass balance and therefore whether it will expand, contract or remain unchanged. Accumulation and ablation do not occur equally over the whole surface of a glacier. Accumulation dominates in the upper regions, where temperature

Figure 3.2   Photograph of the San Rafael Glacier in Chile, a calving glacier. [Photograph: N.F. Glasser]

and precipitation are suitable for snowfall, whereas ablation dominates at the terminus of a glacier, where the climate is relatively warm and where mass can be lost by iceberg calving. It is this spatial imbalance between accumulation and ablation which creates the surface slope that drives ice flow, aided by gravity.

3.2 THE MASS BALANCE GRADIENT: THE GLACIAL DRIVING MECHANISM

Most ablation will occur at the snout or terminus of a glacier, which is usually its point of lowest elevation, where air temperatures will be highest and where iceberg calving may occur. The rate of mass loss (ablation) will decrease with elevation, as atmospheric air temperature falls with altitude. Accumulation may be more uniform over the surface of a glacier, but will tend to increase with elevation. A glacier or ice sheet can, therefore, be divided into two areas: (i) an *accumulation zone*, where accumulation exceeds ablation; and (ii) an *ablation zone*, where ablation exceeds accumulation (Figure 3.3). The line between the accumulation zone and ablation zone is known as the *equilibrium line*; along this line accumulation is balanced by ablation.

On a valley glacier or ice sheet, mass is added at the top in the accumulation zone and taken away through melting and calving at the terminus in the ablation zone. Consequently the surface profile of the glacier will steepen with increasing accumulation (Figure 3.4). In this way the surface slope will steepen until sufficient stress builds up within the ice to cause it to flow. Ice flow transfers mass from

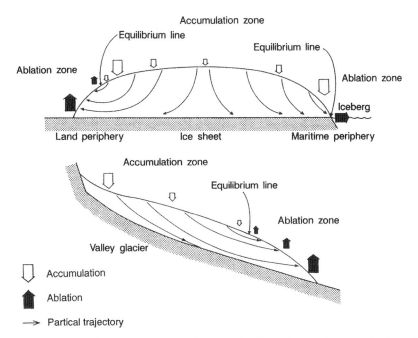

Figure 3.3 Schematic diagram of an ice sheet and valley glacier showing the location of the accumulation zone, the ablation zone and the equilibrium line (the line where accumulation and ablation are equal in any given year). Principal flow paths are also shown. [Modified from: Sugden and John (1976) *Glaciers and Landscape*, Edward Arnold, figure 4.8, p. 63]

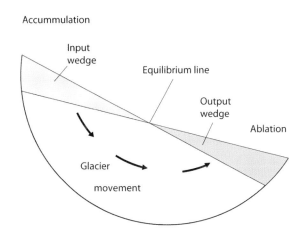

Figure 3.4 Idealised glacier with net accumulation or input 'wedge' and net ablation or output 'wedge'. Glacier flow from the accumulation zone to the ablation zone is necessary if the glacier is to maintain a constant slope. [Modified from: Sugden and John (1976) *Glaciers and Landscape*, Edward Arnold, figure 3.5, p. 41]

the accumulation zone to the ablation zone, thereby reducing the surface slope of the glacier and therefore the stress imposed on the ice. This transfer maintains the glacier slope at a constant or equilibrium angle. It is the gradient therefore between accumulation and ablation across a glacier that causes it to flow. The larger this gradient the greater the glacier flow required to maintain an equilibrium slope. This gradient is known as the *net balance gradient* and is defined as the increase in net balance (accumulation minus ablation) with altitude; that is the sum of the rate of increase in accumulation and the rate of decrease in ablation with altitude. The higher the net balance gradient, the thicker the wedges of accumulation and ablation in Figure 3.4 will be, and as a result the more rapid the glacier flow must be to maintain a constant or equilibrium slope. The net balance gradient will be high on glaciers that have high rates of accumulation and large amounts of ablation. It will therefore be steepest on glaciers that experience warm damp maritime climates and lowest for those in cold dry continental areas (Figure 3.5). Consequently glaciers located in continental areas will flow more slowly than those in warm maritime areas. The rate of glacier flow varies from one glacier to the next and much of this variation is due to differences in the net balance gradient between different glaciers.

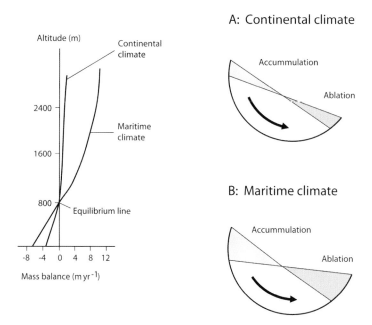

Figure 3.5   The relationship between climate and net mass balance gradient. Two glaciers with net accumulation and ablation 'wedges' are shown: one in a continental climate (A) and one in a maritime climate (B). Although the equilibrium line altitude (ELA) is the same for both glaciers, the maritime glacier has a much steeper mass balance gradient, causing it to flower faster than the continental glacier. [Modified from: Kerr (1993) *Terra Nova*, **5**, figure 3.2, p. 333]

3.3 MECHANISMS OF ICE FLOW

A glacier flows because the ice within it deforms in response to gravity. This gravitational force is derived from the fact that glaciers slope towards their termini as a result of the spatial imbalance between accumulation and ablation discussed in the previous section. If there is no surface slope – no imbalance between accumulation and ablation – the glacier would not flow. The force per unit area set up within a mass of ice by gravity which causes it to deform is known as the *shear stress*. The level of shear stress experienced within an ice mass at any point is dependent upon the ice thickness and the surface slope of the glacier. This can be summarised in the equation:

$$\tau = \rho g (s - z) \sin \alpha$$

where τ is the shear stress at a point within the glacier, ρ is the density of ice, g is the acceleration due to gravity, α is the surface slope of the glacier, s is the surface elevation and z is the elevation of a point within the glacier.

At the base of a glacier the product of $(s - z)$ will be equal to the ice thickness so that the shear stress at the base of a glacier, its *basal shear stress*, is given by:

$$\tau = \rho g h \sin \alpha$$

where τ is shear stress, ρ is the density of ice, g is the acceleration due to gravity, h is the thickness of the glacier and α is the surface slope of the glacier. This equation predicts that basal shear stress will vary with glacier thickness and with surface slope and that high basal shear stress values occur where the ice is both thick and steep.

Different materials can withstand different values of shear stress before they will deform or fracture. In the case of ice, deformation occurs under relatively low values of shear stress and the values obtained from beneath glaciers are remarkably constant. In most situations shear stress at the base of a glacier flowing over bedrock varies between about 50 and 100 kPa (100 kPa = 1.0 bars or kg per cm^2). Ice does not normally deform at levels less than 50 kPa and it cannot usually withstand a shear stress of more than 150 kPa. Glaciers flow because the spatial imbalance between accumulation and ablation leads to an increase in the surface slope, causing the shear stress to increase until the ice deforms or flows.

Values of basal shear stress may be much lower where the glacier flows over a bed that is not rigid but composed of *deformable sediment*. In this situation a large proportion of the forward movement of the glacier may be produced by flow within this deformable sediment as the stress within the glacier is transferred to the bed. Consequently movement may not be controlled primarily by the properties of the ice but by the mechanical and hydrological properties of the sediment below (see Section 3.3.3).

The relatively constant values of shear stress found beneath glaciers flowing over rigid bedrock is an important characteristic and explains why most ice bodies have a parabolic profile; that is, a glacier slope which is steep at the margin and flattens

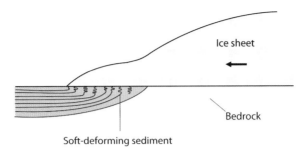

Figure 3.6   Schematic cross-section through an ice sheet to illustrate how an area of deformable subglacial sediment might influence the ice-surface profile. [Modified from: Boulton (1993) in *Holmes' Principles of Physical Geology* (ed. P.McL.D. Duff), Chapman and Hall, figure 20.23, p. 418]

off towards the centre. If shear stress is constant it follows from the equation above that a large ice thickness must be associated with a small surface slope, and a small ice thickness with a large surface slope. The longitudinal profile of a glacier will therefore have a parabolic form, with high slopes at the margin and low slopes in the accumulation zone. The slope of this profile may be reduced or modified if the glacier flows over a layer of deformable sediment (Figure 3.6). This consistency of ice-surface profile makes it possible to reconstruct the form of Cenozoic ice sheets that have long since disappeared if their former margin is known (Box 3.2). The consistency of basal shear stress values is also of importance in providing a way of

---

BOX 3.2: THE PREDICTION OF ICE SHEET PROFILES

It is a reasonable approximation to regard ice as perfectly plastic with a yield stress of 100 kPa: that is, ice will deform plastically if a stress of 100 kPa is applied. Nye (1952) used this assumption to derive a simple equation with which to calculate ice-surface profiles. On a horizontal bed the altitude of the ice surface at any point inland from a known margin can be found from the formula:

$$h = \sqrt{2h_0 s}$$

$$h_0 = \frac{\tau}{\rho g} = 11$$

where h is ice altitude (m), τ is basal shear stress, s is the horizontal distance from the margin (m), ρ is the density of ice and g is acceleration due to gravity.

From comparison with real profiles, it seems that Nye's profile slightly overestimates the slope near the centre of an ice sheet, as shown below.

Despite this, Nye's simple equation has been used widely in reconstructing former ice sheets.

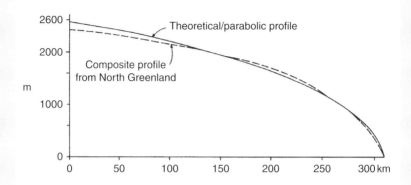

Source: Nye, J.F. (1952) A method of calculating the thickness of ice sheets. *Nature*, **169**, 529–30. [Modified from: Sugden and John (1976) *Glaciers and Landscapes*, Arnold, figure 4.4, p. 60].

critically testing glacier reconstruction based on geomorphological evidence. Former glaciers should obey the same physical laws as glaciers today and should therefore have reconstructed basal shear stress values of between about 50 and 100 kPa over hard substrates (Box 3.3).

BOX 3.3: ESTIMATES OF BASAL SHEAR STRESS AND THE ACCURACY OF GLACIAL RECONSTRUCTIONS

During the last glacial period the Yellowstone National Park in North America was covered by a number of small mountain icefields, one of which is referred to as the Pinedale ice mass. From a careful field study of the landforms and sediments within this area Pierce (1979) was able to reconstruct the morphology of this ice mass. Any reconstruction of a former icefield, on the basis of geomorphological evidence, is likely to be speculative in some respect. Pierce (1979) argued that glaciology offers a method of independently evaluating glacier reconstructions: former glaciers should obey the same physical laws as modern glaciers. In particular the calculation of basal shear stress provides a powerful tool. The two parameters that together specify the shape of a reconstructed glacier – ice thickness and surface slope – are also the primary variables that determine basal shear stress. Empirical observations suggest that the vast majority of modern glaciers, flowing over rigid substrata, have basal shear stresses between 50 and 150 kPa. Pierce (1979) used this to test the validity of his reconstruction of the former Pinedale icefield: if the reconstruction is valid it should give basal shear stress values of this order.

Pierce (1979) calculated the basal shear stress of 50 reaches, each 5–15 km in length, along flow lines within his reconstructed icefield. The results are displayed below on a graph with the ice thickness on one axis and the angle of the former glacier slope on the other. All the values of basal stress fall between 60 and 150 kPa, which is fairly consistent with modern glaciers. The high values of basal shear stress tend to occur in areas likely to have experienced extending flow, whereas the lower values are typical of areas of compressive or decelerating flow. Pierce (1979) concluded that his glacier reconstruction had values of basal shear stress that are both internally consistent and consistent with the values of basal shear stress obtained for modern glaciers. This strongly supports the validity of his reconstruction and illustrates how basic glaciological principles can be used to test the accuracy of glacier reconstructions based on geomorphological evidence.

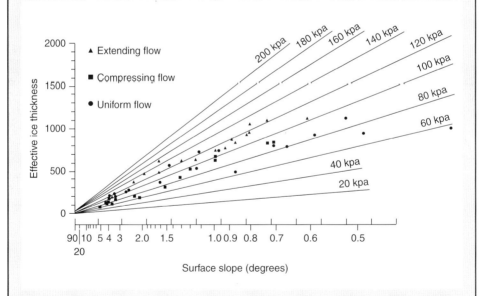

Source: Pierce, K.L. (1979) History and dynamics of glaciation in the Northern Yellowstone National Park area. *US Geological Survey Professional Paper*, 729-F. [Modified from: Pierce (1979) *US Geological Survey Professional Paper*, 729-F, figure 48, p. 71].

Ice can flow in response to the shear stress applied to it through three different mechanisms: (i) by *internal deformation*; (ii) by *basal sliding*; and (iii) by *subglacial bed deformation*.

3.3.1 Internal Deformation

The internal deformation of ice is achieved in two ways: (i) by the process of *creep*; and (ii) by large-scale *folding* and *faulting*.

Creep involves both the deformation of ice crystals and at higher temperatures the mutual displacement of ice crystals relative to one another in response to the shear stress placed upon the ice. The rate of ice creep is a function of the shear stress applied: the greater the shear stress, the greater the rate of ice creep. This relationship is known as *Glen's flow law*. It emphasises the sensitivity of ice creep to shear stress: simply doubling the shear stress will increase the rate of creep by a factor of eight. This explains why most creep takes place in the basal layers of a glacier where shear stress is greatest, because of the greater ice thickness. The rate of deformation is also controlled by temperature, because ice is more plastic at higher temperatures.

The rate of glacier creep may vary down-glacier depending upon whether the glacier is experiencing accelerating (*extensional*) or decelerating (*compressional*) flow. The distribution of zones of extending or compressional flow varies with scale. Compressive flow tends to occur on a glacier where the ice thickness decreases down-glacier (in the ablation zone) and extending flow occurs where ice thickness tends to increase down-glacier (in the accumulation zone). At a small scale, extending flow tends to occur on slopes beneath the ice that steepen down-glacier, whereas compressive flow tends to occur where basal slopes shallow down-glacier. The pattern of surface fractures, *crevasses* (Figure 3.7), on the glacier reflects the type of flow, extending or compressional, that takes place (Figure 3.8). The rate of ice creep is also proportional to temperature. The closer the temperature comes to the melting point the greater the creep rate. For example, in experiments at a fixed stress it was found that the creep rate at $-1°C$ is 1000 times greater than at $-20°C$.

Under certain conditions creep cannot adjust sufficiently fast to the stresses set up within the ice. As a result, *faults* and *folds* may develop. The type of faults that form depend upon whether the ice is experiencing a zone of longitudinal extension or one of compression (Figure 3.8). Areas of compressive flow are also controlled by the temperature of the basal ice as discussed in Section 7.5.

3.3.2 Basal Sliding

There are two main processes by which ice sheets can slide over their beds: (i) enhanced basal creep; and (ii) regelation slip.

Enhanced basal creep is an extension of the normal ice-creep process. It explains how basal ice deforms around irregularities on the ice–bed interface. A glacier bed is not smooth but will contain irregularities, such as bedrock bumps or lodged

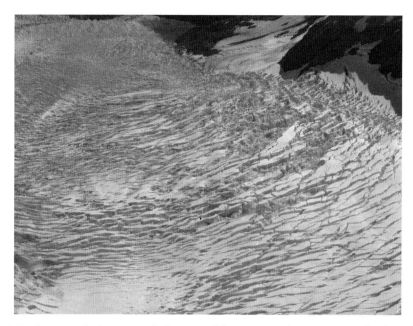

Figure 3.7   Crevasses in the accumulation area of the Tasman Glacier, New Zealand. Ice flow is from bottom right to top left. [Photograph: N.F. Glasser]

boulders, which protrude into the bottom of the moving glacier. Basal ice pressure within the ice increases on the upstream side of such obstacles (see Figure 4.6), and this increases the rate of ice deformation at this point, allowing the ice to flow more efficiently around the obstacle. The larger the obstacle the higher the increase in basal pressure and the greater the rate of deformation. Consequently the process is most efficient for larger obstacles.

Regelation slip occurs when ice at its pressure melting point moves across a series of irregularities or bumps. On the upstream side of each obstacle the basal ice pressure is higher as the ice moves against the obstacle. The melting point of ice falls as pressure rises and as a consequence basal ice melts on the upstream side of obstacles. The meltwater produced will flow around the bump to the downstream side where the pressure is lower and consequently it will refreeze to form *regelation ice*. This process is most effective for small obstacles because the higher temperature gradients across them can drive the heat flux generated by the freezing of the meltwater (release of latent heat) from the downstream side of the obstacle through the rock bump to assist ice melting on the upstream side.

There are, therefore, two processes of basal sliding. One of these, regelation slip, operates best in passing small obstacles, whereas enhanced basal creep works best for larger obstacles. In between the two there is a critical obstacle size range where

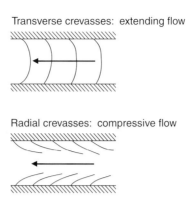

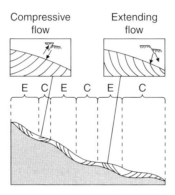

Figure 3.8   Compressive and extending flow in glaciers. Compressive flow is associated with a decrease in subglacial slope angle or a warm-based to cold-based thermal boundary, while extending flow is encountered where the glacier bed steepens or with a cold-based to warm-based thermal boundary. Note that the pattern of surface crevasses differs between the two flow types.

neither process is particularly effective. A bed with obstacles in this size range will, therefore, pose the greatest resistance to basal sliding.

The efficiency of basal sliding depends not only upon the size of obstacles, but also more generally upon the overall level of bed friction. Bed friction is a function of the number of points of contact between the ice and the bed: if these are few then the friction resisting basal sliding will be low. The number of points of contact depends primarily on the amount of water present at the ice–bed interface and its pressure (see Section 4.6). A film of water only a few millimetres thick will reduce friction sufficiently to increase the rate of basal sliding. The presence of water-filled *basal cavities* between the ice–bed interface can also dramatically increase the rate of basal sliding (see Section 4.6). The rate of basal sliding is also affected by the amount of debris within the basal ice; if the debris content is very large the rate of basal sliding may be reduced (see Section 5.1.3).

3.3.3 Subglacial Bed Deformation

When an ice sheet flows over unfrozen sediment it may cause this sediment to deform beneath the weight of the ice. This deformation occurs when the water pressure in the pores or spaces between the sediment grains increases sufficiently to reduce the resistance between individual grains. This allows them to move or flow relative to one another as a slurry-like mass. In response to the shearing force imposed by the overriding glacier this slurry forms a continuously deforming layer on which the glacier moves (Box 3.4). This process can be dramatic. For example, 90% of the forward motion of the Brei∂amerkurjökull in southeast Iceland may be due to subglacial bed or subsole deformation.

BOX 3.4: OBSERVATIONS OF SUBGLACIAL DEFORMATION

Direct observations of deforming sedimentary layers beneath glaciers come from tunnels and cavities. Boulton and Hindmarsh (1987) excavated a series of tunnels within the basal ice of Brei∂amerkurjökull in Iceland, from which a series of strain markers (small cylinders) and other instruments were inserted through drill holes into the subglacial till. The experiment lasted for 136 hours, during which time glacier velocity and the subglacial water pressure were monitored. At the end of the experiment water was pumped from the subglacial bed beneath the ice tunnels, and a section was dug through the bed to study the strain markers. The results are shown in the diagram below. Displacement of the strain markers clearly illustrates the deformation of an upper horizon of till over a relatively undeformed lower layer. It was estimated that approximately 90% of the forward movement of the glacier was due to this subglacial deformation. These observations were used to derive a mathematical flow law to describe the behaviour of glaciers overlying deformable sediments. This flow law has subsequently been used to model and predict the behaviour of deforming sediments beneath glaciers and is critical to theories about the formation of subglacial landforms. More recently, Iverson *et al.* (2007) conducted experiments on a large prism of sediment (1.8 m × 1.6 m × 0.45 m) emplaced beneath 213 m of ice in a tunnel under Engabreen, Norway (see Boxes 5.1 and 5.2). Their experiments lasted between 7 and 12 days, during which time the glacier froze downward into the prism to depths of 50–80 mm, adding sediment to the glacier sole. They also pumped water into the sediment and found that the behaviour of the subglacial sediment was greatly determined by pore-water pressure. At near zero pore-water pressure the glacier slipped across the bed surface; pumping water into the prism increased pore-water pressure, which weakened the sediment and caused it to deform.

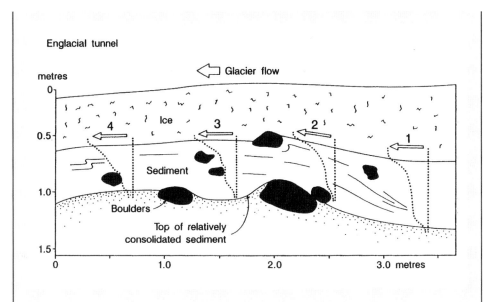

Sources: Boulton, G.S. and Hindmarsh, R.C.A. (1987) Sediment deformation beneath glaciers: rheology and geological consequences. *Journal of Geophysical Research*, **92**, 9059–82. Iverson, N.R., Hooyer, T.S., Fischer, *et al*. (2007). Soft-bed experiments beneath Engabreen, Norway: regelation infiltration, basal slip, and bed deformation. *Journal of Glaciology*, **53**, 323–40. [Modified from: Boulton and Hindmarsh (1987) *Journal of Geophysical Research*, 92, figure 2, p. 9062].

3.4 THE PRINCIPLES OF BASAL THERMAL REGIME

The principal control on which combination of flow processes operate beneath a given glacier – creep or basal sliding – is the temperature of the basal ice. Some glaciers are frozen to their beds and no meltwater is present at the ice–bed interface and basal sliding does not occur (Figure 3.9). Such glaciers are composed of *cold ice*. In contrast, other glaciers are composed of *warm ice*, where basal ice is constantly melting and the ice–bed interface is therefore lubricated with meltwater. In such situations basal sliding is an important component of flow (Figure 3.9). An ice sheet with a warm base therefore has a much greater potential for fast flow and therefore to modify its bed by erosion than one which is frozen to it. Basal ice temperature (*basal thermal regime*) is, therefore, one of the most important controls on the geomorphological impact of a glacier because it controls the pattern of erosion and deposition within it. Not only does basal thermal regime vary between glaciers, but it also varies within a particular ice body.

   The temperature at the base of a glacier is determined by the balance between: (i) the heat generated at the base of the glacier; and (ii) the temperature gradient within the overlying ice, which determines the rate at which the basal heat is

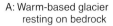

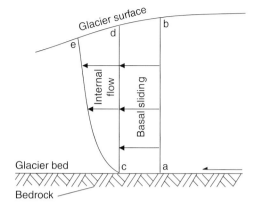

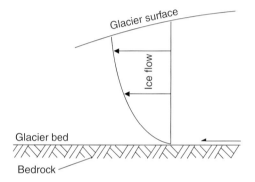

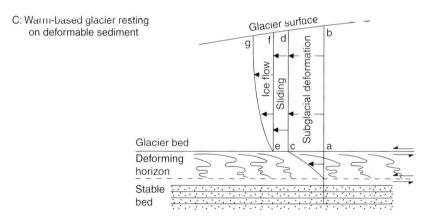

Figure 3.9   The velocity distribution within three glaciers of different basal thermal regimes resting on different substrate. A vertical line a–b inserted in the glacier would be displaced as follows: (A) Warm-based glacier resting on hard bedrock: the vertical line a–b is displaced to c–d by basal sliding and deformed to line c–e by internal deformation. (B) Cold-based glacier resting on hard bedrock: movement is by internal deformation alone. (C) Warm-based glacier resting on deformable sediment: the line a–b is displaced to c–d by subglacial sediment deformation, to e–f by basal sliding and e–g by internal deformation. [Modified from: Boulton (1993) in *Holmes' Principles of Physical Geology* (ed. P.McL.D. Duff), Chapman and Hall, figure 20.20, p. 416]

drawn away by conduction from the ice–bed interface. Heat is generated at the base of a glacier in three ways: (i) by geothermal heat entering the basal ice from the Earth's crust; (ii) by frictional heat produced by sliding at the base of the glacier; and (iii) by frictional heat produced by the internal deformation of the glacier. These three heat sources combine to warm the base of the glacier. The rate at which this heat is conducted away from the base of a glacier depends upon the temperature gradient within the overlying ice. This temperature gradient depends on: (i) the temperature at the base of the ice; (ii) the temperature of the glacier surface; (iii) the thickness of the ice; and (iv) the thermal conductivity of the ice. The temperature of the basal ice is therefore a function of the amount of heat generated at the base of the glacier and the rate at which it is conducted away along the thermal gradient within the overlying ice. Three basal ice conditions can be defined.

1. **Boundary Condition A.** Net basal melting (*warm ice*). In this case more heat is generated at the base of the glacier than can be removed by conduction in the direction of the temperature gradient.

   Basal heat generated > Heat conducted away
   Heat input > Heat output

2. **Boundary Condition B.** Equilibrium between melting and freezing. In this case the heat generated at the base of the glacier is equal to that conducted away along the direction of the temperature gradient.

   Basal heat generated = Heat conducted away
   Heat input = Heat output

3. **Boundary Condition C.** Net basal freezing (*cold ice*). In this case all the heat generated at the base of the glacier is quickly removed from the bed along the direction of the temperature gradient and the ice remains frozen to its bed.

   Basal heat generated < Heat conducted away
   Heat input < Heat output

These three boundary conditions or basal temperature states can be thought of as separate thermal zones, each of which has its own attributes in terms of the types of basal flow (Figure 3.9). Each is also dominated by different processes and geomorphological activity as we will see in later chapters.

In the discussion so far heat transfer within the glacier has been assumed simply to be a function of conduction along the temperature gradient within the ice. In practice heat is also transferred through *advection*. Advection is the transfer of heat energy in a horizontal or vertical direction by the movement of ice or snow. For example, the downward movement of cold snow or firn within the accumulation area of a glacier will lead to glacier cooling. The rate of advection is therefore partly a function of accumulation rates; high accumulation rates result in strong heat fluxes due to the passage of cold ice through the glacier system. Listed below are the principle variables that help determine the basal ice temperature of a glacier.

1. **Ice thickness.** Increasing ice thickness will have the effect of increasing the basal ice temperature. This is due to the insulating effect of ice: more ice equals more insulation.

2. **Accumulation rate.** Basal ice temperature is also affected by accumulation rates through the process of advection. In a glacier advection occurs as the result of the accumulation of fresh snow on the surface gradually moving down through the glacier to the base. If this snow is cold it will cause the glacier to cool and conversely if it is warm it may cause an increase in temperature. For example, in the interior of an ice sheet, where accumulation rates may be very low and where snow accumulates at low temperatures, advection will favour basal freezing. Towards the margin of this ice sheet, where accumulation rates may be higher and where the snow accumulates at higher temperatures, advection may favour basal melting. The incorporation of large amounts of cold dry snow also has the effect of improving the temperature gradient within the ice and therefore the rate of heat conduction, which also helps reduce basal ice temperatures. Advection may also warm a glacier through the upward movement of warm ice in the ablation zone.

3. **Ice surface temperature.** An increase in the surface temperature of a glacier will reduce the temperature gradient within the ice and is therefore likely to increase basal ice temperature. There is a direct relationship between surface temperature and basal temperature: a fall of 1°C in surface temperature causes a fall of 1°C in basal temperature. This may be achieved by incorporating large amounts of wet snow and through summer melting. The percolation of meltwater through an ice body will have the effect of raising its temperature, because as it refreezes latent heat is liberated. For every gram of meltwater that freezes the temperature of 160 g of ice is raised by 1°C.

4. **Geothermal heat.** An increase in the flux of geothermal heat will increase basal ice temperature.

5. **Frictional heat.** An increase in ice velocity will increase the amount of frictional heat generated and in turn increase the basal ice temperature. A glacier flow of 20 m per year produces the same amount of heat as that produced by the average geothermal heat flux. As we will see in Section 3.5 ice flow within a glacier varies spatially, increasing from zero beneath the ice divide of an ice sheet towards the equilibrium line before it decreases towards the ice margin. Consequently the heat generated by friction will also increase to a maximum close to the equilibrium line. This variation in the amount of heat generated by friction is particularly important in determining the spatial variation of ice temperature within a glacier.

The above allows us to predict the situations likely to produce cold- and warm-based ice. *Cold-based glaciers* are likely to occur where the glacier is thin, slow moving and where there is little or no surface melting in summer and the surface layers of ice are cooled severely each winter. A typical temperature profile through a cold glacier is shown in Figure 3.10A. The increase in temperature with depth is due to the insulating effect of the overlying layer of ice and the increase in pressure with depth. The temperature gradient is positive – warmer at the base than at the surface – and therefore the heat will flow from the glacier bed to the surface. Heat will only flow along a negative gradient: from warm to cold. Any heat generated at the base of a glacier will in this case be conducted quickly away from the glacier base through the ice. Boundary Condition C will therefore prevail.

In contrast, *warm-based glaciers* are likely to occur where the ice is thick, fast moving and where summer melting is high. The percolation of meltwater through the glacier body will warm the ice as it refreezes through the release of latent heat and ice temperature may be close to its melting point. The melting point of ice varies with depth due to the change in pressure. Beneath 2000 m of ice, within an ice sheet, melting point will be –1.6°C instead of 0°C. This is termed the *pressure melting point*.

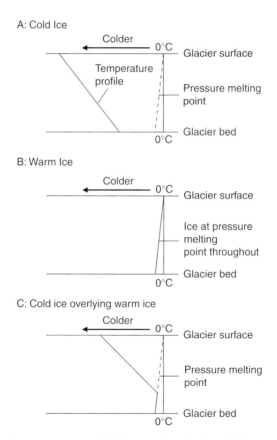

Figure 3.10   Idealised temperature profile through three glaciers. (A) Temperature profile within a cold-based glacier. (B) Temperature profile within a warm-based glacier. (C) Temperature profile where cold ice overlies warm ice at depth. [Modified from: Chorley *et al.* (1984) *Geomorphology*, Methuen, figure 17.4, p. 435]

Within an ice mass close to its melting point the temperature profile will look like that in Figure 3.10B. In this case the profile is positive – colder at the base than at the surface – and consequently any basal heat generated will not be able to escape and Boundary Condition A will therefore apply. The cases outlined above represent two extremes of a continuum. Glaciers in rare instances may be entirely warm- or cold-based, but in practice basal boundary conditions vary both in space and in time within a single glacier. For example, Figure 3.10C shows a situation where cold ice overlies warm ice at the glacier bed.

3.4.1 Spatial Variation of Basal Thermal Regime within a Glacier

Figure 3.11 shows just one way in which the three boundary conditions or thermal zones can be combined within a single glacier. At its simplest this can be thought of as a continuum between cold and warm ice (Figures 3.11A–E). Boundary Condition B has been split into two zones, one in which there is slight net freezing (B^1) and one in which there is slightly more melting (B^2). This reflects the position of this thermal zone within the transition between the two extreme types of basal boundary condition. In Figure 3.11A Boundary Condition C occurs throughout the glacier and the underlying ground is frozen (permafrost); because the glacier is frozen to its bed no basal sliding occurs. In the next diagram the central part of the glacier lies in thermal balance and is neither predominantly melting nor freezing (Figure 3.11B). In this central zone basal sliding will occur. This will cause compression toward the

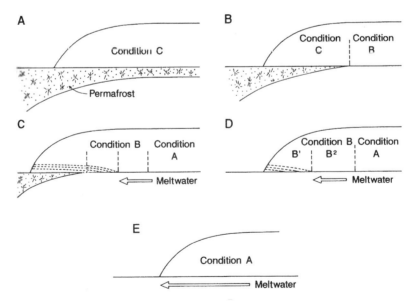

Figure 3.11   The different patterns of basal thermal regime that can exist in a glacier: Condition A, warm; Condition B, thermal equilibrium; Condition C, cold. See text for full explanation of each thermal condition. [Modified from: Boulton (1972) in: *Polar Geomorphology* (eds Price and Sugden), Institute of British Geographers, special publication 4, figure 3.1, p. 4]

transition into the outer zone of the glacier, which is still frozen to its bed and therefore will not experience basal sliding. Situations like this result in large compressive zones near glacier margins in which basal ice is often thrust toward the glacier surface (see Section 7.5). Figure 3.11C shows a more complex pattern of thermal zones, in which the centre of the ice sheet or glacier experiences basal melting. Meltwater passes out from this zone under hydrostatic (water) pressure and then freezes to the bed beneath the cold ice in the intermediate zone (B^1) and new basal ice is formed. Beyond this the glacier is frozen to its bed. Figure 3.11D and E shows the other end of the continuum, moving toward a glacier that is melting at its base and one which therefore will experience basal sliding throughout.

   In practice, the pattern or order of these thermal boundaries within a glacier may vary dramatically. For example, irregular basal topography or climatic variation across a glacier produce more complex localised patterns. Figure 3.12A shows a cross-section through a large mid-latitude ice sheet. The pattern of basal thermal regime within it is one idea of what the pattern may have looked like in one of the mid-latitude ice sheets of the Cenozoic Ice Age. This pattern of basal thermal regime

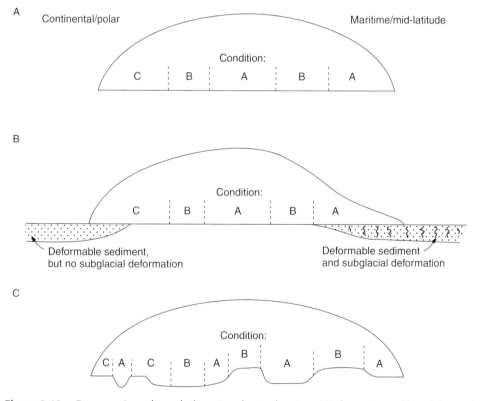

Figure 3.12   Cross-sections through three ice sheets showing: (A) the pattern of basal thermal regime in a schematic ice sheet; (B) the role of the underlying lithology in modifying this pattern; and (C) the role of the underlying topography in modifying this pattern. Condition A, warm; Condition B, thermal equilibrium; Condition C, cold. See text for full explanation of each thermal condition

is based primarily upon the effects of ice thickness and climate and ignores the effects of advection. Climatic variation across the ice sheet results in the contrast between the northern and southern margins: a contrast between a cold continental and a warm maritime climate. Figure 3.12B shows how the presence of a deformable bed can be exploited only where the glacier is warm-based. The pattern of basal thermal regime can be further complicated by introducing basal topography (Figure 3.12C). A deep trench beneath a zone of the ice sheet that would normally be cold at the base may induce melting at the glacier sole. Equally a raised mountain area may experience basal freezing in what would otherwise be a zone of melting (Figure 3.12C).

An alternative view of the basal temperature distribution within the former mid-latitude ice sheets of the Cenozoic Ice Age is obtained when the role of advection is emphasised. In this model the cooling effect of incorporating cold snow, advection cooling, in the ice-sheet centre causes it to be cold-based in the middle. While the increase in frictional heating towards the ice margin and the upward transfer of warm ice by compressive flow (advective warming) in the ablation zone cause the ice sheet to be warm-based towards its margin (Figure 3.13).

Whatever the pattern of basal thermal regime within an ice sheet, it has a profound effect upon the work done by the ice sheet or the geomorphology produced. This point will be returned to in later chapters but is well illustrated

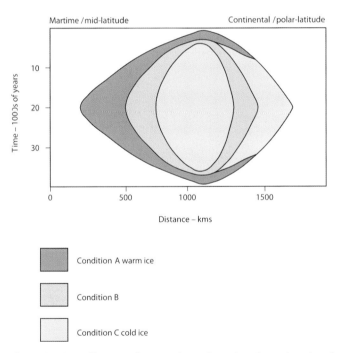

Figure 3.13   Schematic time–distance diagram through an ice sheet showing the growth and decay of the ice sheet and the evolution of its basal thermal regime. As the ice sheet grows and decays, a zone of warm-based ice migrates across the landscape. The pattern of basal thermal regime at the ice-sheet maximum can be envisaged by taking a time-slice through the ice sheet at approximately 20 000 years

by the control of basal regime on the mechanisms of ice flow (Figure 3.9). Where the ice is frozen to its bed, irrespective of the nature of the substrate, flow may occur only by internal deformation and movement is concentrated above the bed. As a consequence the potential to modify the bed is small. In contrast, in zones of thermal melting flow may occur by basal sliding and by subglacial deformation where the substrate is appropriate (Figure 3.9) and the ice sheet has a more profound effect on its bed.

3.4.2 Temporal Variation in Basal Thermal Regime within a Glacier

Just as the basal ice temperature may vary within a glacier it may also vary through time. In particular the pattern of basal thermal regime within an ice sheet is not static. As an ice sheet grows and decays the pattern of basal ice temperature within it will also evolve. Consequently the pattern of processes controlled by it will also vary through time as the pattern of thermal regime changes. Figure 3.13 shows a hypothetical cross-section through a mid-latitude ice sheet, showing the evolution of the pattern of basal thermal regime within it. This diagram is highly schematic and in reality the patterns are likely to have been much more complex. It serves to illustrate, however, that at any location the temperature of the ice above may change as the ice sheet grows and decays and consequently the processes operating upon it will also change. This pattern results primarily from two main factors: (i) the increase in frictional heating towards the equilibrium line maintains a warm outer ring to the ice sheet; and (ii) cooling by advection causes the central part of the ice sheet to remain cold.

At a much smaller scale the pattern of basal thermal regime, and in particular the boundaries between one thermal zone and another, may change as a result of local or regional fluctuations in: (i) ice velocity; (ii) accumulation; (iii) geothermal heat flux; and (iv) ice thickness. On a longer time scale glacial erosion may modify basal topography, causing variations in the thermal regime. Any of these variables may change the temperature of the basal ice and therefore the dynamic nature of the processes operating at the ice–bed interface.

3.5 PATTERNS AND RATES OF ICE FLOW

Within a glacier flow usually follows the direction of the surface slope. Figure 3.3 shows a cross-section through both an ice sheet and a glacier. In an ice sheet the ice flows in two opposite directions from the summit or *ice divide* (Figure 3.3). In the accumulation area, flow takes place downwards into the ice, counteracting the upward growth of the surface through accumulation. In the ablation zone the surface is lowered by ablation, which causes ice to effectively rise towards the surface. Ice may also rise towards the surface in the ablation zone due to compressional flow at the glacier margin. Flow within a valley glacier or channel shows a similar pattern. Ice flow is maximum in the centre of the channel or valley and near the surface: the point furthest from the frictional resistance of the valley sides (Figure 3.14).

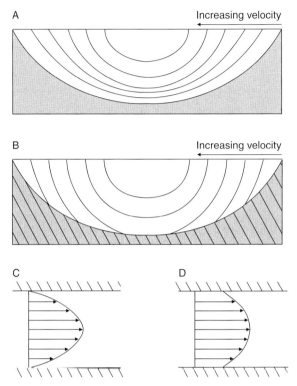

Figure 3.14   The pattern of ice flow in a glacierised valley. (A) The flow pattern in cross-section where no basal sliding is present. (B) The flow pattern in cross-section where basal sliding is present. (C) The flow pattern in plan view where no basal sliding is present. (D) The flow pattern in plan view where basal sliding is present. [Modified from: Summerfield (1990) *Global Geomorphology*, Longman, figure 11.5, p. 265]

In an ideal ice sheet the rate of flow will tend to increase from the ice divide towards the equilibrium line, where it will reach a maximum, before decreasing towards the terminus. This observation can be explained schematically with reference to Figure 3.15. If we assume a constant average velocity through the whole thickness of the ice sheet, then at a point at a given distance (x) from the ice divide the horizontal flow velocity (u) must be sufficient to remove by flow all the accumulation (a) which occurs up-ice of that point. The amount of ice or discharge that can pass through this point is given by:

$$discharge = uh$$

where u is the average ice flow velocity (m s^{-1}) and h is ice thickness (m).

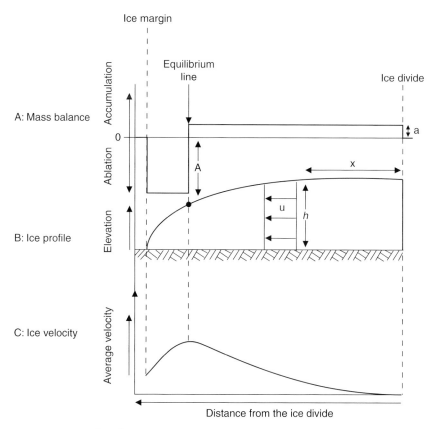

Figure 3.15   Pattern of ice flow within an ice sheet. (A) A simplified mass balance pattern on the ice surface. (B) The ice sheet in profile. (C) Ice-flow velocity within the ice sheet. Velocity rises from zero at the ice divide to a maximum near the equilibrium line, after which it declines. [Modified from: Boulton (1993) in: *Holmes' Principles Of Physical Geology* (ed. P.McL.D. Duff), Chapman and Hall, figure 20.14, p. 412]

This discharge must equal the accumulation rate up-ice of that point if the ice sheet is to maintain a steady state (i.e. size). Therefore the amount of ice to be discharged by the glacier is given by:

$$discharge = xa$$

where x is the distance from point x to the ice divide (m) and a is the average accumulation (m yr^{-1})

It follows therefore that average ice velocity (u) can be calculated from:

$$uh = discharge = xa$$

This can be rearranged to give:

$$u = \frac{xa}{h}$$

where u is the average ice-flow velocity (m s^{-1}), h is ice thickness (m), x is the distance from point x to the ice divide (m) and a is the average accumulation (m yr^{-1}).

Using this simple equation, ice-flow velocity increases from zero at the ice divide ($x = 0$) to a maximum at the equilibrium line, thereafter it decreases. Clearly the assumptions made in calculating the velocity pattern in Figure 3.15 are considerable, but the pattern produced holds as a general model or first approximation of

Figure 3.16   A surge-type glacier, Lowell Glacier in the Yukon. Note the folded and looped medial moraines on the ice surface. [Photograph: M.J. Hambrey]

the velocity pattern within an ideal ice sheet. This pattern is significant because it implies that: (i) little or no geomorphological work will take place beneath ice divides due to low ice velocities; and (ii) that most geomorphological work will be done beneath the equilibrium line, which is usually located relatively close to the ice margin.

Most glaciers have velocities in the range of 3–300 m per year, but their velocity can reach 1–2 km per year in steep terrain or where there is a high mass balance gradient. A few glaciers flow at speeds that are much higher. These are commonly associated with large outlet glaciers from ice sheets such as that in Greenland or Antarctica. In these the flow of *ice streams* is channelled down valleys and velocities may reach as much as 7–12 km per year. These ice streams may drain significant areas of an ice sheet and because they are fed by a large accumulation area their velocities are not normally limited by the supply of ice (i.e. the rate of accumulation).

Some glaciers may also experience periodic *surges* in ice flow, often 10–100 times greater than previous ice velocities (Figure 3.16). Surges are usually limited by the amount of ice available in the accumulation zone so that increased flow rates cannot be sustained. Not all glaciers are prone to surges and those that do, appear to surge at regular intervals. It has been suggested that only about 4% of all glaciers surge, although they tend to be concentrated in certain geographical areas, such as in Svalbard and Alaska.

It has been suggested that there are two modes of glacier flow: one 'normal' and one 'fast' (Figure 3.17). In an ice stream the 'fast' flow mode is maintained because of the availability of ice within the large accumulation zone of the ice

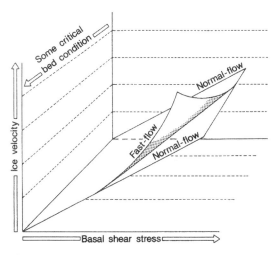

Figure 3.17 Modes of glacier flow. Two states exist: 'fast flow' and 'normal flow'. The existence of 'fast flow' is dependent on some critical bed condition (e.g., changes in the subglacial hydrological system or thermal regime), without which it cannot occur. A surge-type glacier is able to switch between fast and normal flow rates through time.

sheet that it drains. In the case of surging valley glaciers the pulse of 'fast' flow is limited by the amount of ice available in the much smaller accumulation zone. Once this ice has been discharged the flow must return to 'normal'. It is possible to conceptualise a surge as the product of an excess of accumulation of ice above that which 'normal' flow can discharge. This excess accumulation may be stored in the accumulation area until it reaches a critical level, when it may trigger a pulse of 'fast' flow. The periodic nature of a surge is explained in this model by the time necessary between pulses of 'fast' flow to build up the excess of ice or the stress necessary to trigger the event. This will vary from one glacier to next, which explains why different glaciers surge with different periodicities. The exact nature of the instability that generates a surge and the mechanisms of 'fast' flow are not well understood at present (Box 3.5: see also Section 4.5).

BOX 3.5: DO SURGE-TYPE GLACIERS DEFORM THE SEDIMENT BENEATH THEM?

Despite a number of years of research it is still unclear exactly how soft-bedded glaciers interact with the sediment beneath them. In an attempt to resolve this, Truffer *et al.* (2000) conducted a series of experiments at Black Rapids Glacier, a 40 km long, surge-type glacier in the central Alaska Range. They drilled to the bed of the glacier at a spot where the measured surface velocities are high and seasonal and annual velocity variations are large. The drilling revealed a layer of subglacial sediment ('till'), up to 7 m thick, under the glacier. The sediment was water saturated. A string of instruments, containing three tiltmeters (instruments designed to measure very small changes from the horizontal) and a piezometer (a small diameter water-well designed to measure the hydraulic head of groundwater in aquifers), was lowered into the sediment. The tiltmeters monitored the inclination of the borehole at the ice-sediment interface for 410 days at depths of 1 m and 2 m in the sediment. They showed that no significant deformation occurred in the upper 2 m of the sediment and that no significant amount of basal motion was accomplished by sliding of ice over the sediment. The measured surface velocity at the drill site was about 60 m yr^{-1}, of which 20–30 m yr^{-1} can be accounted for by ice deformation. The inference is that almost the entire amount of basal motion, 30–40 m yr^{-1}, was taken up at a depth of greater than 2 m in the sediment, possibly in discrete shear layers, or as sliding of sediment over the underlying bedrock. Truffer *et al.* (2000) concluded that the large-scale mobilisation of sediment layers such as this could be a key factor in initiating glacier surges.

Source: Truffer, M., Harrison, W.D. and Echelmeyer, K.A. (2000) Glacier motion dominated by processes deep in underlying till. *Journal of Glaciology*, **46**, 213–21.

3.6 GLACIER RESPONSE TO CLIMATE CHANGE

The size of a glacier is determined primarily by climate. Changes in climate will cause its margins to expand or contract, because climate controls a glacier's mass balance (Box 3.6). If the balance is positive (accumulation greater than ablation) it will expand and grow. If it is negative then the glacier will contract. Glaciers are constantly responding to changes in their mass balance budget, adjusting to annual variations as well as to long-term trends (Figure 3.18).

BOX 3.6: RECENT CHANGES IN THE MASS BALANCE OF ARCTIC GLACIERS

Net surface mass balance is the sum of winter accumulation and summer losses of mass from glaciers and ice sheets. Where this is positive, glaciers grow and where it is negative, they recede. Net surface mass balance is determined primarily by changes in climate. Dowdeswell *et al.* (1997) examined records of surface mass balance for more than 40 Arctic ice caps and glaciers going back in some cases as far as the 1940s. They deliberately excluded the large Greenland Ice Sheet because its surface mass balance is so poorly constrained. The remaining Arctic glaciers and ice caps cover about 275 000 km^2 of the archipelagos of the Canadian, Norwegian, and Russian High Arctic and the area north of about 60°N in Alaska, Iceland and Scandinavia. The records show that most Arctic glaciers have experienced predominantly negative net surface mass balance over the past few decades. Dowdeswell *et al.* (1997) found no uniform trend in mass balance for the entire Arctic, although some regional trends occur. Examples are the increasingly negative mass balances for northern Alaska, due to higher summer temperatures, and increasingly positive mass balances for maritime parts of Scandinavia and Iceland, due to increased winter precipitation. The negative mass balance of most Arctic glaciers may be a response to a step-like warming in the early twentieth century at the termination of the cold 'Little Ice Age'. Dowdeswell *et al.* (1997) calculated that the Arctic ice masses outside Greenland are at present contributing about 0.13 mm yr^{-1} to global sea-level rise.

Source: Dowdeswell, J.A., Hagen, J.O., Björnsson, H., *et al.* (1997) The mass balance of circum-Arctic glaciers and recent climate change. *Quaternary Research*, **48**, 1–14.

To understand how the glacier system responds to climate change we need to consider what happens when a glacier thins or thickens due to changes in its mass balance. If climate deteriorates – temperatures fall and/or snowfall increases – every part of the glacier is likely to thicken. This thickening will result from either an increase in accumulation or a reduction in ablation and may cause the ice margin

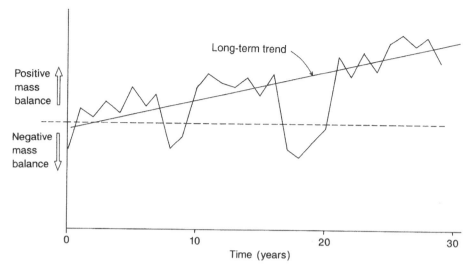

Figure 3.18   Short-term and long-term trends in mass balance

to advance. Conversely, a climatic warming will lead to overall thinning of the glacier and retreat of its margin. Glacier response to such changes in mass balance may be either stable or unstable. A stable response is one in which the glacier thickens or thins in proportion to the size of the change in mass balance. An unstable response is one in which the glacier thickens or thins out of proportion with the size of the change in mass balance that triggers it. The type of response depends on whether the flow is extending or compressive (Figure 3.8). Unstable behaviour occurs more often with compressive flow. This can be illustrated in relation to Figure 3.19 where A–A and B–B are two cross-sections in a glacier subject to compressive flow. The discharge (Q volume per unit time) of ice through cross-section B–B is less than that through cross section A–A by the amount of ablation (ab) which occurs on the glacier surface between points A and B ($Q_{AA} = Q_{BB} + ab$). If we now place a uniform layer of snow over the glacier surface it can be shown mathematically that the increase in discharge caused by the layer of snow is proportional to the original flow through the section. Consequently, because the flow through cross-section A–A is larger than that through B–B, the increase in discharge due to the new layer of snow will be greater at A–A than at B–B ($Q_{AA} > Q_{BB} + ab$). Therefore the ice must thicken between point A and B in order to accommodate this increased volume. With continued flow this effect will be accentuated down-ice and the glacier will progressively thicken towards its snout, eventually causing it to advance.

Changes in mass balance are propagated down-glacier by means of *kinematic waves*. A kinematic wave can be viewed as a bulge in the glacier surface. The greater thickness of ice in the bulge will locally increase the basal shear stress and therefore the rate of ice deformation, and consequently the glacier velocity. The bulge will therefore move down ice faster than the thinner ice on the up-ice and down-ice side

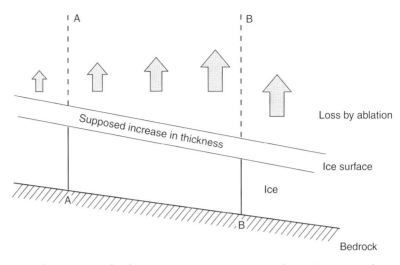

Figure 3.19   The response of a glacier to an increase in accumulation in an area of compressive flow. See text for details. [Modified from: Sugden and John (1976) *Glaciers and Landscape*, Edward Arnold, figure 3.12, p. 48]

of the bulge. It is important to stress that it is the bulge that moves and not the ice: a boulder on the surface would temporally move at a higher velocity when it was on top of the bulge and then return back to its normal velocity as the bulge passed by. Kinematic waves originate in the vicinity of the equilibrium line. In general the equilibrium line represents the junction between predominantly extending flow in the accumulation zone and the compressive flow typical of the ablation zone. An increase in ice thickness, caused by a deterioration in climate, will therefore cause a stable thickening of the glacier above the equilibrium line and an unstable response below it. It is this difference in thickening response that initiates a kinematic wave. The wave, once formed, travels down the glacier at a rate faster than the ice velocity. When the wave reaches the snout, often years later, it may initiate an advance of the ice margin. The passage of a kinematic wave is similar to the passage of a flood wave within a river. It is important to emphasise that kinematic waves are rarely visible at the glacier surface. They are simply the mechanism by which mass balance changes are propagated throughout the glacier system: the means by which it adjusts to changes in climate by extension or contraction of the ice margin.

The rate at which a kinematic wave passes through a glacier is highly variable. Some glaciers respond quickly to changes in mass balance whereas others do not. The length of time for glacier adjustment to a change in mass balance, the *response time*, depends on the sensitivity of the particular glacier to change and upon the nature of the change. For example, an excess of ablation at the glacier snout one year may cause a glacier snout to retreat almost immediately, by contrast an advance in the snout due to an excess of accumulation must first feed through the whole glacier before it has an effect. The response time is also controlled by the sensitivity of the glacial system, which depends on a variety of parameters, including glacier morphology and activity (i.e. the mass balance gradient). At a simple level the larger the ice body the slower the response time. Large ice sheets will respond only to large and sustained changes in mass balance, whereas small cirque or valley glaciers will often respond quickly to minor fluctuations. The rate of response of these smaller ice bodies is influenced by their morphology, as illustrated in Figure 3.20. Here two valley glaciers with different slopes are affected differently when the equilibrium line is raised by an identical amount. The ratio of accumulation to ablation area of Valley Glacier B is much more sensitive than that of Valley Glacier A and therefore will be more responsive to climatic change. In a similar manner glaciers which are very active, that is, they have large mass balance gradients, will respond much more quickly to changes in mass balance than those that are less active.

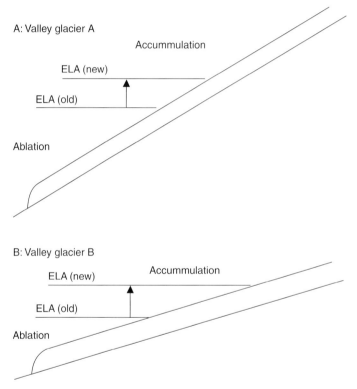

Figure 3.20   The impact of a rise in equilibrium line altitude (ELA) on two valley glaciers with different gradients. The impact of a change in the ELA is greater on valley glacier B because it has a shallower gradient and the rise in ELA therefore affects a larger surface area. [Modified from: Kerr (1993) *Terra Nova*, **5**, figure 4, p. 335]

The link between climate, mass balance and glacier response is complex and may not always be apparent. This is particularly the case where glaciers terminate or calve into water. On land a glacier can react to changes in input by extending or withdrawing its snout. An extension of the snout means that an increased surface area is exposed to ablation because the glacier advances into warmer areas at lower altitude. A glacier in a deep *fjord* – a glacial valley that has been drowned by the sea – cannot do this so easily. This can be illustrated by considering a tidewater glacier in an ideal fjord of constant width and depth (Figure 3.21). The ablation area of the glacier is limited to the lower reaches of the glacier and, more importantly, to the amount of ice that can be melted or discharged as icebergs from the cross-sectional area of the snout. If there is a shift to a positive net balance the glacier will begin to advance. As it cannot extend to lower altitudes to enhance ablation, it will continue to advance until it can spread out and increase the cross-sectional area exposed to melting and calving. This will occur only at the fjord mouth or at a point at which the fjord widens or deepens. Consequently, fjord glaciers are particularly sensitive to changes in mass balance, and relatively minor climatic changes can cause spectacular variations in the position of snouts. The most stable positions for calving glaciers within fjords include: (i) fjord mouths; (ii) fjord bifurcations; (iii) points where fjords widen, narrow or are bordered by low ground; and (iv) where they deepen or shallow. These locations are known as *pinning points*, points where the ablation geometry of the ice margin may be changed and control the location of calving ice margins (Figure 3.21).

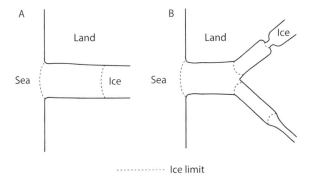

Figure 3.21   The role of fjord geometry in controlling the location of ice margins. (A) In a fjord of constant width and depth a glacier can advance to the fjord mouth, but will not advance beyond the fjord mouth because of the increase in calving rate at this point. (B) Examples of topographic pinning points (dashed lines) for a calving glacier. Calving glaciers will tend to be stable at these pinning points

   Understanding the interaction of climate, mass balance and glacier response is important not only in examining the response of glaciers to relatively minor climatic fluctuations but also to the more dramatic and long-term fluctuations associated with the onset of a glacial cycle. The growth and decay of large ice sheets is a complex problem. Traditionally, ice sheets are considered to grow through a sequence of larger and larger ice bodies – snow patches>cirque glaciers>valley glaciers>icefields>ice caps>ice sheets – developing first in upland areas and then expanding into lowland regions as the ice bodies merge and grow. More recently computer models have suggested that the sequence of growth may follow a slightly different pattern – snow patches → cirque glaciers → valley glaciers → piedmont lobes → small ice sheet. In this scenario valley glaciers first develop in mountainous areas and flow out into low-lying areas where the ice spreads out as large lobes, known as *piedmont lobes*. These lobes merge and thicken rapidly in an unstable fashion to produce small ice sheets, which then merge to produce larger ones. The piedmont lobes thicken and grow dramatically because they have low gradients and consequently low ice velocities, and therefore cannot discharge the ice pouring into them from the fast flowing valley glaciers that drain the steep mountainous areas behind. Topography may also play a very important part in the rate at which an ice sheet develops. This has been illustrated by a computer model of a former ice cap that existed in the Scottish Highlands at the close of the last glacial cycle, during a cold period known as the Younger Dryas (10 000 years ago). This computer model illustrated the sensitivity of this ice cap to the mountain topography of the highlands: certain types of topography acceler-ated its growth. As illustrated in Figure 3.22 for a given deterioration in climate, parts of the ice cap that advanced into large basins grew more dramatically than

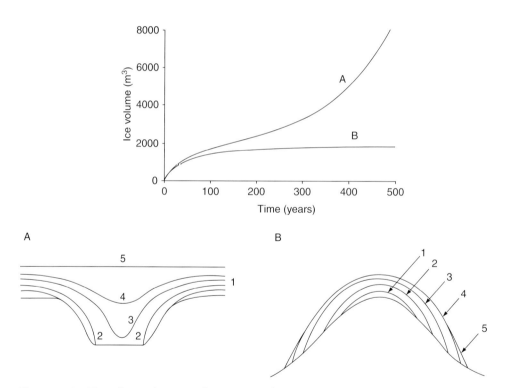

Figure 3.22   The effects of topography on rates of ice-sheet growth in a numerical ice-sheet model. The top panel shows ice-sheet volume through time, and situations (A) and (B) represent ice advancing into a topographic basin (A) and ice advancing down a mountain slope (B). Situation (A) is represents rapid non-linear growth in which a small change in climate can have a dramatic effect on ice volume, while situation (B) represents stable growth. [Modified from: Payne and Sugden (1990) *Earth Surface Processes and Landforms*, **15**, figure 6, p. 632]

those centred on topographic ridges. Consequently the location of large topographic basins have a dramatic effect on the rate at which ice sheets grow. In these locations small deteriorations in climate may have a dramatic effect on the size of the ice body.

Once established an ice sheet will continue to grow provided that there is an excess of precipitation over that which the ice sheet can discharge and ablate. Growth will be driven by a variety of positive feedback systems. For example, the Earth's atmosphere is characterised by strong altitudinal gradients in temperature and precipitation. As an ice sheet grows an increasing proportion of its area will lie at more favourable altitudes for accumulation, facilitating further growth. Ice sheets will grow until: (i) their size is limited by the available space, such as the edge of the continental shelf and the presence of deep water; (ii) precipitation starvation sets in when the interior of an ice sheet becomes so removed from sources of precipitation around its margins that its rate of accumulation falls; and (iii) climate changes, reducing accumulation or increasing ablation. Theoretically ice sheets should reach an optimum or equilibrium size for the prevailing climate and topographic location. The ice sheet will exist until some change in this climate occurs to cause it to decay (*deglaciation*).

Deglaciation may be driven by either a decrease in precipitation and therefore accumulation or an increase in ablation or a combination of both. An increase in ablation may be achieved not only through a rise in air temperature but also through a rise in sea level. Rising sea level may increase the area subject to calving and therefore induce rapid ablation. Rapid calving will increase the discharge of ice towards these margins, a process that may initiate the development of large fast-flowing ice streams draining the ice-sheet centre. It has been suggested that this process may cause the rapid 'draw-down' and deglaciation of an ice sheet. Traditionally there are two models used to explain how deglaciation occurs.

1. **Catastrophic down-wasting by areal stagnation.** In this model the equilibrium line rises quickly above the ice sheet, depriving it of accumulation. As a consequence down-wasting is rapid and traditionally believed to be associated with little forward flow. This model has been used to explain the rapid decay of mid-latitude ice sheets during the Cenozoic 'Ice Age' and the presence of extensive areas of stagnation-type landforms within the limits of these former ice sheets.

2. **Active deglaciation by regular ice-marginal retreat.** In this model ice sheets decay in a regular fashion through the contraction and retreat of the ice margin. Forward flow is maintained even as the margin of the glacier retreats. This mechanism can occur both slowly and quickly.

Although widely postulated as a mechanism of glacier decay, areal stagnation is now considered by many geologists to be of only regional or local importance. It is unlikely that ice sheets will be completely deprived of accumulation during decay. More significantly, ablation will always be maximum at the margin and minimum at the ice divide, maintaining the surface ice-sheet gradient, basal shear stress and therefore forward flow. Stagnation on a local or regional scale may occur, however, where

4: Glacier Hydrology

This chapter explains the characteristics and significance of glacier hydrology and its links to glacier motion. We examine sources of glacier meltwater, how water is stored in and moves through glaciers, the concepts of subglacial water pressure and the processes of glacial meltwater erosion.

4.1 GLACIER HYDROLOGY

Glacier hydrology is the study of water flow through glaciers. An understanding of glacier hydrology is important for a number of reasons.

1. On many glaciers meltwater is the main ablation product and runoff from these glaciers is an important economic asset in many parts of the world, providing water for drinking and sanitation, irrigation for crops and hydroelectric power. For example, an estimated 500 million people depend on the tributaries of the glacier-fed Indus and Ganges rivers for irrigation and drinking water.

2. Glacier hydrology is an important control on the dynamics of glaciers and ice sheets. Water flow through glaciers is intimately related to glacier dynamics through glacier motion, for example through enhanced glacier sliding (see Section 2.2) and because subglacial meltwater is required for subglacial sediment deformation.

3. Where water is stored in or beside a glacier, its sudden release can constitute a glacier hazard (see Box 2.3). Rapid drainage of moraine-dammed or ice-dammed lakes can threaten lives and infrastructure downstream.

4. Glacial meltwater removes debris from the ice–rock interface and carries it beyond the confines of the glacier, where it is deposited. In many formerly glaciated parts of the world the distribution of these glacial meltwater sediments (e.g., sand and gravel) forms a valuable economic asset. Sand and gravel deposits can also be important water-bearing aquifers.

Glacial Geology: Ice Sheets and Landforms Second Edition Matthew R. Bennett and Neil F. Glasser
© 2009 John Wiley & Sons, Ltd

4.2 SOURCES OF GLACIAL MELTWATER

Glacial meltwater is derived from the melting of ice in one of three positions: *supraglacial*, which means on the ice surface, *subglacial*, at the bed, and *englacial*, which means within the glacier (Figure 4.1). Melting occurs whenever there is sufficient heat to turn the ice back into water, and this heat can be supplied by: (i) solar radiation; (ii) friction generated by ice flow; and (iii) heat derived from the Earth's crust beneath the glacier (geothermal heat). Melting on the ice surface is the most important source of glacial meltwater on many glaciers. On temperate glaciers it is often measured in metres per year, whereas englacial and subglacial melting may contribute only millimetres per year. Ice-surface melting is of course highly

Figure 4.1   Examples of glacial meltwater drainage. (A) Supraglacial stream on the surface of Austre Brøggerbreen in Svalbard. (B) Moulin on the surface of Austre Brøggerbreen in Svalbard. (C) Englacial tunnel or conduit melting out of stagnant ice in front of Fox Glacier, New Zealand. Note the rounded and subrounded material melting out of the conduit. (D) Subglacial tunnel and meltwater emerging from the snout of Fox Glacier, New Zealand. [Photographs: N. F. Glasser]

seasonal because surface melting by solar radiation generally requires positive air temperatures (Figure 4.2). The volume of water within the meltwater system also depends on the amount contributed by rainfall, snowmelt on the glacier surface and from valley-side streams, all of which can add significant quantities of externally derived water.

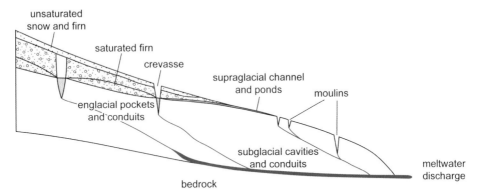

Figure 4.2   Sources of meltwater and principal transfer routes in a typical temperate alpine glacier. In the accumulation zone, water percolates down through the snow and firn to form a perched water layer on top of the nearly impermeable ice and then flows from the from perched water layer into crevasses. In the ablation zone, once all the seasonal snow has melted, water flows directly across the glacier surface into crevasses and moulins. [Modified from: Fountain and Walder (1998) *Reviews of Geophysics*, **36**, figure 1, p. 300]

4.3 STORAGE OF WATER IN GLACIERS

It is important to note from the outset that meltwater not only flows through glaciers but that it can also be stored within a glacier in a number of ways. Storage occurs as ice, snow and water at a number of different spatial and temporal scales. For example, in subglacial settings, meltwater can be stored in cavities, in the pore space of subglacial sediments and in subglacial lakes. It can also be stored within the ice in englacial water pockets, tunnels and cavities, and on the glacier surface as snow and firn, and in supraglacial lakes. Water can also be stored adjacent to a glacier in proglacial and ice-marginal lakes.

Water storage in glaciers occurs at three time-scales (Figure 4.3). *Long-term storage* (years to centuries and longer) occurs as glacier ice and firn. This storage affects the long-term water balance of glacierised catchments and has the potential to influence global sea level. *Intermediate-term storage* (days to years) includes the seasonal storage and release of snow and water. Intermediate-term storage affects the runoff characteristics of glacierised catchments and their downstream river-flow regimes. *Short-term storage* (hours to days) includes the daily effects of drainage through a glacier including meltwater routing through snow and firn, as well as englacial and subglacial pathways. In addition to these time-dependent

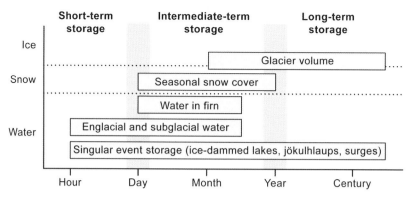

Figure 4.3   Schematic graph showing different forms of glacier storage and their corresponding time-scales. [Modified from: Jansson *et al.* (2003), *Journal of Hydrology*, **282**, figure 1, p. 117]

processes there are also *event-driven storage releases*, including floods released by glacier surges and the drainage of subglacial, moraine-dammed or glacier-dammed water bodies (see Box 2.3).

4.4 METHODS OF STUDYING GLACIER HYDROLOGY

Direct investigation of subglacial drainage networks and channel systems is very difficult so we rely on a number of indirect methods to study water flow through glaciers and ice sheets. Five main techniques can be used.

1. In *dye-tracing experiments*, fluorescent dye is injected into the drainage network at the glacier surface via moulins and crevasses, or below the surface via boreholes. The dye mixes with water flowing in the glacier, and its emergence at the glacier snout is recorded using a fluorometer (an instrument that measures the level of fluorescence of water). The speed and space–time pattern of dye return enable inferences to be made concerning the structure of the drainage network. Dye travelling quickly through the glacier retains its identity as a well-defined packet with high dye concentration, indicating an *efficient drainage network*. Dye travelling slowly through the glacier loses its identity because it has more chance to become dispersed, indicating an *inefficient drainage network* with water storage within the glacier. Dye-tracing experiments can also be used to trace water transport paths from individual moulins or crevasses to the snout thereby building a picture of the drainage system.

2. Studies of *meltwater quality*, the sum of chemical properties such as dissolved ions, micro-organisms and suspended sediment concentration, also provide information about the configuration and dynamics of subglacial drainage systems. For example, the flux of suspended sediment is related to the changing properties of the subglacial drainage network. Subglacial drainage reorganisation and flood events tend to sweep large parts of the glacier bed free of sediment; thereafter the

discharge of suspended sediment tends to fall because the supply of sediments is exhausted. Similarly, the chemical signature of meltwater leaving the glacier reflects the different type and speed of chemical reactions taking place under the glacier. The spatial extent and speed of water flow are important; water that flows slowly across large areas of the glacier bed tends to pick up large quantities of solutes, whereas water that flows quickly across limited areas of the glacier bed has little chance to pick up dissolved material. As the chemical signature in meltwater is strongly controlled by these properties of the drainage system, it is possible to distinguish between glacial meltwater that has been in contact with the bed and meltwater that has not.

3. *Borehole studies* require a hole to be drilled into the ice from the glacier surface using a hot-water drill. This allows direct measurement of parameters such as subglacial water pressure by monitoring fluctuations in the level of the water in boreholes. In this way, it is possible to follow the changing time–space pattern of water pressure, and use this to infer changes in the behaviour of the subglacial drainage system. Cameras and down-borehole tele-viewers can also be used to look at englacial and subglacial ice structures and their relationship to water flow.

4. *Radar radio-echo sounding* and *ground-penetrating radar* (GPR) can be used to infer the water content of a glacier because of the differences in radar-wave velocities caused by the dielectric properties of water, air, sediment and ice. This technique has been used to map the outline of englacial and subglacial drainage channels on glaciers, and to relate their distribution to the subsurface hydrology and to glacier thermal and structural conditions.

5. *Former glacier beds* can be used to infer patterns of subglacial drainage by mapping the distribution of landforms related to meltwater flow and reconstructing the former drainage pathways that they represent. Mapped landforms include the location of former cavities, ice-abraded areas, chemically altered areas, precipitate-filled depressions and meltwater channels. This technique works particularly well on areas of hard bedrock in front of receding glaciers where meltwater landforms are most likely to be preserved (Box 4.1).

BOX 4.1: DETERMINING THE GEOMETRY OF FORMER SUBGLACIAL DRAINAGE SYSTEMS

The geometry of former subglacial drainage systems can be determined by mapping bedrock surfaces that previously lay directly beneath a glacier. Walder and Hallet (1979) mapped landforms and sediments on limestone bedrock exposed by recession of Blackfoot Glacier, a small cirque glacier in Glacier National Park, Montana, USA. They mapped the following features, deemed to reflect different degrees of modification of the bed by glacial abrasion, water erosion, chemical dissolution and chemical precipitation.

Chemically altered areas. These are distinguished by the presence of abundant, furrowed, subglacially precipitated calcite deposits on the down-glacier sides of bedrock protuberances. They indicate close ice–rock contact beneath the former glacier, separated by a thin water film.

Abraded areas. These are distinguished by the scarcity of both subglacially precipitated calcite deposits and solutional furrows, with abundant striations. They represent areas of intimate ice–rock contact.

Cavities. Areas of steeply inclined bedrock, usually down-glacier of bedrock highs or breaks in slope that exhibit extensive solutional features and lack of evidence of ice contact. Some of these features also have a thin, discontinuous coating of subglacial precipitate around their peripheries. These areas are interpreted as the sites of former water-filled subglacial cavities.

Precipitate-filled depressions. Shallow (<50 mm) bedrock depressions nearly filled with subglacial precipitate. These features are interpreted as representing former shallow subglacial cavities.

Channels. Narrow, elongate depressions, typically 50–250 mm deep, 100–200 mm wide and 2–5 m long, with well-developed solutional features. Many have a thin discontinuous coating of subglacial precipitate. They are interpreted as subglacial meltwater channels.

Karst. Narrow, deep, linear to sublinear bedrock depressions formed by dissolution along two joint sets. They tend to interconnect with underground passages, and are therefore interpreted as glacially modified subaerial karst features.

Walder and Hallet (1979) concluded that a near-continuous, non-arborescent network of cavities acted as the primary meltwater drainage system beneath the glacier. They estimated that nearly 20% of the former glacier sole was separated from the bed by water-filled cavities and that the remainder of the glacier–rock interface was characterised by a thin water film. The ideas presented by Walder and Hallet (1979) have been developed numerically by a subsequent study at Glacier de Tsanfleuron in Switzerland (Sharp *et al.*, 1989).

Sources: Sharp, M., Gemmell, J.C. and Tison, J.-L. (1989). Structure and stability of the former subglacial drainage system of the Glacier de Tsanfleuron, Switzerland. *Earth Surface Processes and Landforms*, **14**, 119–34.Walder, J.S. and Hallet, B. (1979). Geometry of former subglacial water channels and cavities. *Journal of Glaciology*, **23**, 335–46.

4.5 GLACIER HYDROLOGICAL SYSTEMS

Glacial meltwater finds its way through the glacier from its point of origin along a variety of different flow paths (Figure 4.2). Meltwater derived from surface melting and from basal melting will tend to follow different paths through a glacier, although both will invariably involve channel flow either within or on the glacier. The type of channel system within a particular glacier depends primarily on its thermal regime. On glaciers consisting of cold ice, meltwater is unable to penetrate without freezing and tends to be confined to surface and ice-marginal channels. On glaciers consisting of warm ice, meltwater is able to penetrate without freezing and water flow can occur in supraglacial, englacial and subglacial channels. In glaciers with mixed thermal regimes, more complex flow patterns are likely and water can flow through supraglacial, englacial and subglacial channels. It can also be stored in small subglacial lakes at the junction between ice of different temperatures (see Section 4.3).

4.5.1 Supraglacial and Englacial Water Flow

Supraglacial channels tend to be less than a few metres wide and may exploit structural weaknesses within the ice (Figure 4.1A). In plan they may adopt either meandering or straight courses. Velocities within these channels are usually high because their smooth sides offer little frictional resistance. On temperate glaciers, supraglacial channels are usually short and are interrupted by *crevasses* or vertical shafts known as *moulins* (Figure 4.1B), which divert the water into the glacier where they may become *englacial* (Figure 4.1C).

The internal geometry of moulins and englacial drainage routes has been studied in detail on Storglaciären in Sweden and on debris-covered Himalayan glaciers. Both sets of drainage routes develop by exploiting crevasses and other pre-existing structures within the ice. Detailed three-dimensional mapping of englacial drainage conduits on the Himalayan glaciers shows that their location is determined by pre-existing lines of high hydraulic conductivity and that channels are graded to local base level. Grading is accomplished by headward retreat and downcutting, producing canyon-like passages within the ice. The Storglaciären moulins are also structurally controlled. When a crevasse opens on the glacier surface it often intersects a supraglacial meltwater stream. The crevasse will then fill with water, until it opens and deepens sufficiently to intersect englacial drainage passages, at which point the water drains away. When glacier flow moves the crevasse into an area of compression the crevasse may close, but the heat carried into it by the meltwater may keep the drainage channel open and thereby form a moulin. The Storglaciären moulins consist of near-vertical shafts, 30–40 m deep, which feed englacial tunnels that descend from the base of the moulin at angles of between 0° and 45°. The orientation of the englacial tunnels is often controlled by the orientation of the original crevasse from which the moulin formed.

4.5.2 Subglacial Water Flow

Subglacial drainage systems can take one of four main configurations.

1. **Configuration 1.** Water can flow in a thin film with a thickness of millimetres between the glacier and its bed, often known as a *Weertman film*. In this configuration, the meltwater is spread widely across the bed. A Weertman film is most

likely to develop where meltwater is derived primarily from basal melting and in situations where there are restricted inputs of surface meltwater. Mathematical analyses suggest that a water film is unstable because the film tends to rapidly organise itself into networks of small channels. The presence of a thin film of meltwater is central to theories of *glacier sliding* because it acts as a lubricant between the ice and its substrate (see Section 3.3.2).

2. **Configuration 2.** *Conduit* or *tunnel networks* discharge meltwater through a small number of large channels (Figure 4.1D). Tunnel networks cover a limited area of the glacier bed and are efficient in transferring meltwater through the glacier. These channels can be within the ice (englacial), or at the ice–bed interface (subglacial). Where channels at the glacier bed cut upwards into the ice they are termed *Röthlisberger channels* or *R-channels*. In plan view, this type of drainage network is often dendritic. The size of the tunnels is determined by the balance between the processes that act to enlarge them (i.e. ice melt of tunnel walls from flowing water) and processes that act to close them (i.e. ice deformation). The natural shape for a tunnel or conduit, well away from the bed, is near-circular due to these two opposing factors (Figure 4.1D). More importantly, the size of a conduit will vary with changes in discharge over a matter of weeks. As melt-water discharge increases, so the conduits grow in size.

3. **Configuration 3.** *Linked-cavity systems* are found where water collects in low-pressure areas within the cavities that are created as ice slides across bedrock obstacles. These cavities cover much of the glacier bed, and are connected by a tortuous network of small links to create a more-or-less continuous drainage network (Figure 4.4). The links can be cut down into bedrock (*Nye* or *N-channels*; Figure 4.5), or can be cut upwards into ice as small R-channels. In this drainage configuration, meltwater is spread across a wide area of the bed and, because the channel geometry is relatively inefficient, meltwater transit times are slow.

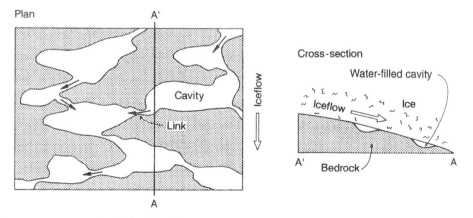

Figure 4.4   Network of linked basal cavities in plan and cross-section. Each cavity is linked by N- or R-channels. [Reproduced with permission from: Hooke (1989) *Arctic and Alpine Research,* **21**, figure 4, p. 226]

Figure 4.5   A Nye channel on Isle of Mull, western Scotland. [Photograph: N. F. Glasser]

4. **Configuration 4.** Meltwater can also flow within subglacial sediments if sufficient soft, potentially deformable sediments exist beneath the glacier. To be
effective, this style of drainage requires the presence of a layer of sediment at
the base of the ice, unlike the other three drainage styles, which relate to 'hard'
rock beds. Subglacial sediments are usually permeable, so that they deform
easily when saturated and subjected to stresses transmitted to the bed by the
overlying glacier. The precise mechanism by which the water moves through the
sediment is unclear. Suggested mechanisms include: advection the water within
the sediment layer is carried forward as the sediment deforms; *Darcian flow*,
whereby water moves through the pore spaces of the sediment from areas of
high water pressure to low water pressure under the influence of the hydraulic
potential gradient; within pipes and small channels in the subglacial sediment
itself; or as a thin film or sheet of meltwater at the upper surface of the sediment.

An important distinction in subglacial hydrology is between distributed and discrete drainage systems. Configuration 1 (Weertman film) is often referred to as a *distributed drainage system* because the meltwater is widely distributed across the glacier bed. Configurations 2 and 3 are often referred to as *discrete drainage systems* because the meltwater is confined to discrete channels and tunnels at the glacier bed. Configuration 4 (meltwater flow within subglacial sediments) can be regarded as either a distributed or discrete drainage system, depending on whether the water flows through the sediment as Darcian flow (distributed), or in a sheet (distributed) or in tunnels and pipes in the sediment (discrete).

It is important to remember that the drainage system in many glaciers is constantly undergoing both spatial and temporal changes. Consequently, the drainage system beneath an individual glacier can vary in space; for example if the glacier bed is 'patchy' and composed of a discontinuous sediment cover of variable thickness over bedrock. It may also vary through time; for example seasonally, where there is a well-documented transition from an early melt-season within an inefficient drainage system to a more efficient drainage system later in the melt season. Rapid transitions between different subglacial drainage configurations are also possible where water flow is non-steady-state, for example beneath surge-type glaciers (see Section 3.5).

4.6 SUBGLACIAL WATER PRESSURE

4.6.1 Subglacial Water Pressure and Effective Normal Pressure

Subglacial water pressure has an important role in many subglacial processes, through its control on the effective normal pressure beneath a glacier. *Effective normal pressure* is the force per unit area imposed vertically by a glacier on its bed. For a cold-based glacier it is effectively equal to the weight of the overlying ice; thick ice imposes a greater pressure than thin ice. This is summarised by:

$$N = pgh$$

where N is the normal effective pressure, p is the density of ice, g is acceleration due to gravity and h is ice thickness.

If water is present at the glacier bed, however, the effective normal pressure is reduced by an amount equal to the subglacial water pressure. Put crudely, the greater the water pressure the more it can support the weight of the glacier and thereby reduce the effective normal pressure acting on the bed. The equation is modified to:

$$N = pgh - wp$$

where N is the normal effective pressure, p is the density of ice, g is acceleration due to gravity, h is ice thickness and wp is the subglacial water pressure.

This is only true where the glacier has a flat bed. Effective normal pressure is modified by the flow of ice over obstacles (Figure 4.6). As ice flows against the upstream side of an obstacle the effective normal pressure increases by an amount proportional to the rate of glacier flow against the obstacle. Effective normal pressure is also reduced in the lee or on the downstream side of the obstacle (Figure 4.6). The pressure fluctuation caused by the flow of ice against the obstacle is, therefore, positive on the upstream side and negative on the downstream side. The negative pressure fluctuation on the downstream side of an obstacle may cause

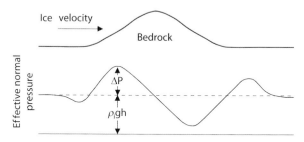

Figure 4.6   Schematic diagram of the distribution of effective normal pressure at the glacier bed as it flows over a bedrock obstacle. [Modified from: Boulton (1974), in *Glacial Geomorphology* (ed. Coates), George Allen and Unwin, figure 8, p. 55]

a cavity to form in its lee if the effective normal pressure at this point is exceeded (Figure 4.7). *Cavity formation* is favoured by: (i) thin ice (ii) high basal water pressures, which reduce effective normal pressure; and (iii) high rates of basal sliding, which produce large pressure fluctuations over obstacles. Theoretical calculations show that cavities can open at sliding velocities of about 9 m per year beneath a thickness of 100 m of ice, whereas velocities of 35 m per year are required with ice thickness of the order of 400 m.

Basal water pressure is controlled by four variables: (i) glacier thickness – the greater the weight of the overlying ice the greater the water pressure; (ii) the rate of water supply – the input of large amounts of meltwater may increase the pressure; (iii) the rate of meltwater discharge – an efficient subglacial drainage system will reduce water pressure; and (iv) the nature of the underlying geology – permeable bedrock, for example, will reduce water pressure. Variations in the rate of water supply and the rate of meltwater discharge are responsible for much of the seasonal variation in water pressure present at some glaciers. Early in the melt season water pressure may be very high due to the abundance of meltwater and the relative inefficiency of the channel network (see Section 4.7). As the subglacial channel network develops during the ablation season discharge becomes more efficient and the water pressure generally falls.

As we will see in Chapters 5 and 6, variations in water pressure and its influence on effective normal pressure and cavity formation are very important for the processes of glacial erosion. Basal water pressure is also important in determining the rate of basal sliding (see Section 3.3.2). Effective normal pressure helps determine the friction experienced between a glacier and its bed. If the water pressure

Figure 4.7   Photograph of a large subglacial cavity in the lee of a rock step beneath the southern margin of San Rafael Glacier in Chile. Ice flow is from right to left over the rock step. [Photograph: N. F. Glasser]

rises, effective normal pressure will fall, thereby reducing basal friction and consequently increasing basal sliding. This explains why sliding velocity often increases during the summer melt season or after a large rainfall event. For example, field observations on the Unteraargletscher have shown that it moves vertically by 0.4 m at the start of the melt season due to increased water pressure. This is followed by a similar downward movement at a constant rate over the next 3 months. The glacier velocity increases significantly when the surface is raised. Variations in basal water pressure have also been linked to glacier surges. For example, the surge of the Variegated Glacier in Alaska, during 1982–1983 is believed to have been triggered by a change in the subglacial drainage system. Prior to the surge the glacier had a subglacial drainage system dominated by a few large tunnels. This appears, however, to have changed to a system dominated by linked subglacial cavities in which the water pressure rose dramatically due to the lower rate of discharge possible from such a system. This rise in water pressure facilitating rapid glacier flow during the surge, but at the end of the surge this stored water was released as a large flood and the subglacial system reverted to a large integrated tunnel system. The cause of this change in drainage system is unclear, but is believed to be central to the rapid glacier flow of this surge.

In summary therefore variation in basal water pressure has an important role to play in determining the flow dynamics of a glacier and is also important in the processes of glacial erosion (see Chapter 5).

4.6.2 Water Pressure Gradients

The orientation of this network of conduits and tunnels is controlled by the water pressure gradient within the glacier. Water will flow down the pressure gradient from areas of high to low pressure. It is possible to determine the nature of this pressure gradient within a glacier and therefore the direction of water flow within it. Figure 4.8 shows a hypothetical water-filled tube beneath a glacier. The weight of ice above point A is equal to the weight of the water column B–C which it forces up. A line between A and C defines a surface of equal potential pressure. Along this line the pressure due to the weight of the overlying ice is equal to the water pressure it generates. If we now move the tube towards the right, closer to the ice margin, the

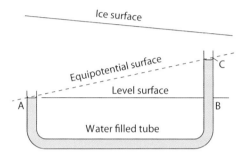

Figure 4.8   Diagram to illustrate the hydraulic head which drives water flow within a glacier. The weight of the ice above point A is equal to the elevation of the water column BC. The thinner the ice above point A, the less the hydraulic head. Consequently the hydraulic head or potential will fall towards the ice margin or in the direction of glacier slope. Water flows from areas of high hydraulic potential to areas of low hydraulic potential.

weight of the ice above point A will fall and consequently the water column B–C will be lower. A new lower equipotential surface is defined. Water will flow at right angles to these equipotential surfaces from a surface of higher potential pressure to one of lower potential pressure. As a consequence englacial conduits and tunnels will be orientated perpendicular to surfaces of equipotential pressure (Figure 4.9). Some moulins may be an exception to this, reflecting their origin as crevasses. The geometry of the equipotential surfaces within a glacier is determined by the variation in ice thickness, which is controlled primarily by the surface slope of the glacier and secondarily by the slope of the underlying topography. The surface of a glacier does not always slope in sympathy with the slope of the glacier bed. As a consequence subglacial meltwater may not always flow directly down the maximum slope beneath the glacier and may in some cases even flow uphill. Under an ice sheet water flow will be approximately radial, in sympathy with the surface slope and the direction of ice flow, but will deviate around hills and bumps and be concentrated in topographic depressions such as valleys.

It is possible to calculate the water pressure potential at a series of points at the base of a glacier from knowledge of the variation in ice thickness. These points can be contoured to define a surface known as the *subglacial hydraulic potential surface* (Figure 4.9). Provided that any subglacial tunnel is completely water filled then the tunnel should be orientated at right angles to this hydraulic surface. The ability to calculate this surface is a useful tool in the

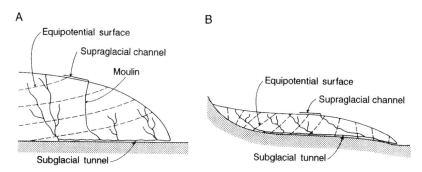

Figure 4.9   The pattern of equipotential surfaces within a glacier (i.e. surfaces of equal hydraulic potential). Water will always flow from areas of high hydraulic potential to areas of low hydraulic potential, and therefore it will flow at right angles to the equipotential surfaces as depicted here.

interpretation of the glacial landform record (Box 4.2). If the subglacial tunnel is not, however, completely full of water, something which may occur at the ice margin, then the water flow and the orientation of the tunnel may be controlled by the underlying topography beneath the glacier and not by the subglacial water pressure surface. The presence of gravity driven subglacial flow at the margin of glaciers has been investigated using *dye-tracing experiments* (Box 4.3). This technique can also be used to understand the seasonal evolution of sub-glacial drainage systems (Box 4.4).

BOX 4.2: EQUIPOTENTIAL SURFACES: A TOOL IN GLACIAL RECONSTRUCTION

There is a long-running controversy about the nature of the glacial history of the area of Antarctica known as the Dry Valleys in Victoria Land. Under contemporary conditions the largest valleys contain no glacier ice, although small valley glaciers are present in some of the tributary valleys. Some researchers argue that this area was once covered by part of the Antarctic Ice Sheet, whereas others have suggested that it was only ever covered by expanded local valley glaciers. Sugden *et al.* (1991) set out to resolve this debate. Within the Dry Valleys there are a series of spectacular subglacial meltwater channels (see Box 6.3). In order to resolve whether these channels were formed under an extension of the Antarctic Ice Sheet or beneath local ice Sugden *et al.* (1991) calculated the subglacial hydraulic potential surface which would have existed within the area had it been covered by an ice sheet. The subglacial meltwater channels should, if formed beneath the ice sheet, run at right angles to the contours used to define this surface (i.e. they should run down the pressure gradient).

This involved first estimating the location of the ice margin on the assumption that the area had been covered by an ice sheet, and then estimating the ice surface profile. The ice-surface morphology was estimated using the profile equation of Nye (1952) (Box 3.2), which assumes a parabolic ice-surface profile. This ice surface was then compared with the topography of the Dry Valleys in order to determine variations in ice thickness from which the subglacial hydraulic potential surface was calculated using a simple equation developed by R. L. Shreve. This equation states that:

$$\emptyset = \emptyset_o + p + p_w g h$$

where $\emptyset$ is the subglacial water pressure potential, $\emptyset_o$ is a constant, p is water pressure in a conduit at the base of the glacier, which is assumed to be equal to the pressure of the overlying ice, p_w is water density, g is acceleration due to gravity and h is ice thickness.

If the subglacial meltwater channels were formed beneath the ice sheet then they should run at right angles to the subglacial equipotential surface calculated from Shreve's equation. If they had formed beneath local ice the pattern would be different, reflecting the equipotential surface within a local valley glacier. The meltwater channels were found to run at right angles to the equipotential contours calculated for the ice sheet, which provides strong support for the contention that the area was once covered by an extension of the Antarctic Ice Sheet. This example illustrates how the application of simple principles in glaciology can resolve problems of interpretation in areas of former glaciation.

Source: Sugden, D.E., Denton, D.H. and Marchant, D.R. (1991). Subglacial meltwater channel system and ice sheet over riding of the Asgard range, Antarctica. *Geografiska Annaler*, **73A**, 109–21.

BOX 4.3: DETERMINING THE GEOMETRY OF MODERN GLACIAL DRAINAGE SYSTEMS

The geometry of the englacial and subglacial drainage systems of modern glaciers can be examined using a technique known as *dye-tracing*. Dye is added to supraglacial meltwater as it disappears into the glacier. The time taken for this meltwater to emerge at the glacier snout is recorded. If this procedure is repeated at a large number of locations over a small valley glacier it is possible to obtain an idea of its internal drainage structure.

Sharp *et al.* (1993) did this for the Haut Glacier D'Arolla in Switzerland. A total of 342 dye injection experiments were conducted using 47 moulins distributed widely across the glacier surface. These experiments were combined with detailed measurements of both the surface and subsurface topography of the glacier using conventional field survey techniques and radio-echo sounding. The dye injection experiments were used to reconstruct the pattern of internal drainage within the glacier. This was then compared with the basal water pressure gradients within the glacier calculated from R. L. Shreve's equation (Box 4.2). The drainage system should flow normal to the contours that define the subglacial hydraulic potential surface. A close approximation was found between the reconstruction and the pattern of contours or water pressure gradient. Some discrepancies were, however, noted, which might imply that the drainage close to the ice margin was at least seasonally driven by the slope of the subglacial topography and not by the surface slope of the glacier.

Source: Sharp, M., Richards, K., Willis, I., et al. (1993). Geometry, bed topography and drainage system structure of the Haut Glacier D'Arolla, Switzerland. *Earth Surface Processes and Landforms*, **18**, 557–71.

BOX 4.4: SEASONAL CHANGES IN THE MORPHOLOGY OF SUBGLACIAL DRAINAGE SYSTEMS

Nienow *et al.* (1998) used dye-tracing techniques to investigate the glacier-wide pattern of change in the englacial and subglacial drainage system of the 4 km-long Haut Glacier d'Arolla in Switzerland during the ablation seasons of 1990 and 1991. Their data indicate that over the course of a melt season the drainage system changed from an initially inefficient system to a much more efficient system of large subglacial channels and tunnels. The efficient drainage system gradually grew at the expense of the inefficient system until, by the end of the melt season, the subglacial channel system extended at least 3.3 km from the snout of the glacier and drained nearly all of the supraglacially derived meltwater passing through the glacier. The upglacier limit of the subglacial channel system closely followed the retreating snowline as it migrated up-glacier, with rates of headward channel growth reaching $\sim 65\,$m a day. From these data, Nienow *et al.* (1998) inferred that the removal of snow (with its high albedo and significant water storage capacity) from the glacier surface resulted in a dramatic increase in the volume of supraglacial runoff into moulins. Daily runoff cycles also became more pronounced as the snowline receded up-glacier. High water pressures within the distributed drainage system caused it to develop rapidly into a channelised system. This study provides a neat example of how a subglacial drainage system evolves beneath a valley glacier during the course of the ablation season.

Source: Nienow, P., Sharp, M. and Willis, I. (1998). Seasonal changes in the morphology of the subglacial drainage system, Haut Glacier d'Arolla, Switzerland. *Earth Surface Processes and Landforms*, **23**, 825–43.

4.7 DISCHARGE FLUCTUATIONS

The discharge of meltwater from glaciers varies dramatically both on a diurnal (daily) and seasonal basis. Diurnal discharge variations reflect atmospheric air temperatures and therefore the pattern of daily ablation on a glacier. Discharge is usually low in the early morning and rises in the late afternoon or evening (Figure 4.10A). This diurnal fluctuation is suppressed in winter, but increases towards the late summer when the rate of daily ablation reaches its maximum. Seasonal fluctuations are equally dramatic (Figure 4.10B). They reflect two factors: (i) the seasonal nature of ablation; and (ii) the seasonal development of the internal drainage network within warm-based glaciers. We can develop a simple model for the drainage pattern for glaciers in a strongly seasonal climate.

1. **Spring melt.** On the glacier ablation of winter snow begins in the spring and as a consequence water pressures within the glacier begin to increase. In front of the glacier, ice in proglacial rivers breaks -up and melting of winter snow proceeds rapidly.
2. **Late spring melt.** Ablation of winter snowfall is well advanced on the glacier surface. Discharge in all channels, conduits and tunnels increases. The conduits grow in size and the internal drainage network within the glacier develops. The amount of water available within the glacier exceeds that which can be discharged by the internal drainage network. Discharge from the glacier into the proglacial channel system steadily increases.
3. **Early summer.** With the development of a well connected drainage network within the glacier much of the stored water is released. The daily discharge exceeds the amount of daily melt on the glacier surface and in the course of a few weeks the drainage system discharges the vast majority of its total annual discharge (*nival flood*). This increase in discharge may be associated with a sudden rise in glacier velocity, the so-called 'spring event' (Box 4.5).
4. **Late summer.** The drainage network within the glacier has reached its optimum efficiency. All the stored water within the glacier has been discharged. Daily discharge matches the amount of melt achieved each day. Water pressures within the glacier are usually at their minimum.

A

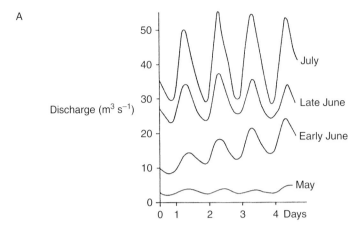

B

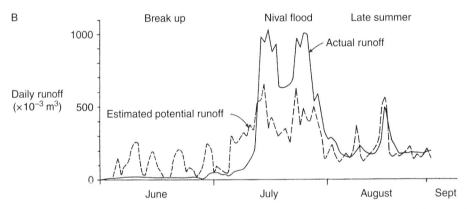

Figure 4.10   Schematic discharge fluctuations for alpine glacial meltwater streams. (A) Diurnal (daily) fluctuations, these become more pronounced as the melt season proceeds. (B) Seasonal fluctuations. The difference between estimated potential runoff and actual runoff reflects the efficiency of the subglacial drainage network. At the start of the melt season the channel and tunnel network is poorly developed and therefore actual runoff is less than potential (i.e. water is stored in the glacier). As the melt season proceeds, the drainage network grows in efficiency so that actual and potential runoffs become similar. [Modified from: Elliston, G.R (1973), *Symposium on the hydrology of glaciers*, International Association of Scientific Hydrology, **45**, figure 1, p. 80]

BOX 4.5: LINKS BETWEEN SUBGLACIAL HYDROLOGY AND GLACIER VELOCITY

Many valley glaciers experience an early-melt-season high-velocity event (the so-called 'spring event'), where glacier velocities increase suddenly. Working on Haut Glacier d'Arolla, Switzerland, Mair *et al.* (2003) investigated the relationship between the spring event and the nature of the subglacial hydro-logical drainage system. They collected data concerning the spatial patterns of ice-surface velocity, internal ice deformation rates, the spatial extent of high subglacial water pressures and rates of subglacial sediment deformation around the time of three spring events. During two of the three events, their data suggest widespread ice–bed decoupling, particularly along the axis of a known subglacial drainage channel, which explains the observed increase in glacier velocity. The other event was marked by less extensive ice–bed decoupling and sliding along the drainage axis, suggesting subglacial sediment deformation may have been important. This link between subglacial hydrology and glacier velocity has also been established for the polythermal John Evans Glacier, Ellesmere Island, Canada (Copland *et al.*, 2003). These spring events demonstrate that changes in subglacial hydrology are intimately linked to changes in glacier dynamics. The overall conclusion of these studies is that there is a close association between: (i) the timing and spatial distribution of the temporal pattern of surface water input to a glacier; (ii) the formation, seasonal evolution and distribution of subglacial drainage pathways; and (iii) horizontal and vertical glacier velocities. Consequently increases in glacier velocity can be directly attributed to changes in glacier hydrology.

Sources: Copland, L., Sharp, M.J. and Nienow, P.W. (2003). Links between short-term velocity variations and the subglacial hydrology of a predominantly cold polythermal glacier. *Journal of Glaciology*, **49**, 337–48. Mair, D., Willis, I., Fischer, U.H., *et al.* (2003). Hydrological controls on patterns of surface, internal and basal motion during three 'spring events': Haut Glacier d'Arolla, Switzerland. *Journal of Glaciology*, **49**, 555–67.

5. **Autumn.** Cessation of melting on the glacier causes a dramatic drop in discharge. The internal network of conduits and tunnels within the glacier begins to collapse as the water flowing within them declines and they close due to ice deformation. Only a few major drainage arteries contain sufficient flow to be maintained.

6. **Winter.** The degree to which the drainage network shuts down depends on the climate and severity of the winter. Most meltwater discharge if there is any will be derived from internal melting.

The degree to which the internal drainage network of conduits and tunnels within a glacier collapses each year is probably highly variable. Some shut down will occur in all cases, but in large ice sheets the major discharge arteries are likely to be maintained because the rate of flow, due to melt generated by internal deformation and geothermal heat, is likely to be much larger. The smaller and thinner the glacier, the more likely it is that the drainage system evolves each year.

On some glaciers this pattern of diurnal and seasonal discharge fluctuation are interrupted by catastrophic subglacial floods known as *jökulhlaups*. These are high-magnitude events, often several orders of magnitude greater than normal peak flows. They are distinct from glacial lake outburst floods (GLOFs; see

Box 2.3) in that they are more restricted in scope being wholly ice derived, however in practice the distinction is rather artificial. Jökulhlaups may occur in one of two ways: (i) through subglacial volcanic activity; and (ii) through the drainage of ice-dammed lakes. Volcanic activity beneath glaciers is common today in Iceland. A spectacular example is Grìmsvötn, beneath the Vatnajökull ice cap. Here water builds up subglacially above a volcano, which then drains catastrophically in the space of a few hours, through a 50 km long subglacial tunnel. The average volume of water discharged from Grìmsvötn is between 3 and 3.5 km^3. The volcano has erupted, causing major jökulhlaups, in 1960, 1965, 1972, 1976, 1982, 1983, 1986, 1991, 1996, 1998 and 2004. During the last major eruption in 1996, the jökulhlaup discharged approximately 3 km^3 of water with a peak flow rate of 45 000 m^3 s^{-1}. This is more than the discharge of the Mississippi River.

Jökulhlaups due to the sudden drainage of ice-dammed lakes are much more widespread. Ice-dammed lakes occur wherever a glacier blocks the down-valley flow of water (Figure 4.11). When it occurs, the outburst from such lakes is typically catastrophic in nature. Lakes can drain catastrophically for a number of reasons: (i) vertical release of the ice dam by flotation caused by rising water levels in the lake; (ii) overflowing of an ice dam, causing rapid melting of the dam due to friction from the water flow; (iii) destruction or fissuring of an ice dam by earthquakes; and (iv) enlargement of pre-existing tunnels beneath the dam by increased water flow and melt-enlargement due to frictional heating. In many situations the maximum water

Figure 4.11   Photograph of a partially drained ice-dammed lake in South West Greenland. Note the stranded icebergs that indicate the lake level has dropped during a drainage event. [Photograph: N. F. Glasser]

level in a lake is limited by a spillway or channel via which the lake can drain. In these cases some other mechanism is required to trigger catastrophic lake drainage.

The shape of the jökulhlaup hydrograph will be determined by the nature of the trigger and the volume of water that is drained (Figure 4.12). Jökulhlaups induced by volcanic activity tend to produce high-magnitude, short-duration floods, whereas jökulhlaups caused by the drainage of ice-dammed lakes may produce floods with a greater duration and lower magnitude (Figure 4.12). As we might expect, the size of the flood peak from a jökulhlaup caused by the drainage of ice-dammed lakes is proportional to the volume of water within the lake (Box 4.6). The high-magnitude nature of jökulhlaups makes them of considerable geomorphological significance (see Section 8.2).

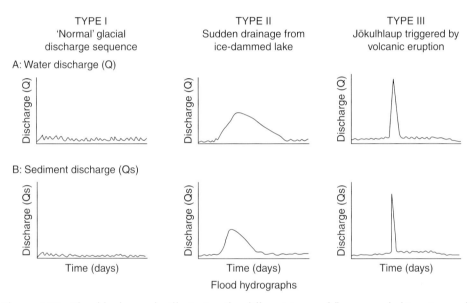

Figure 4.12 Flood hydrographs illustrating the different types of flow recorded in a 'normal' glacial river, during drainage of an ice-dammed lake and during a jökulhlaup. [Modified from: Maizels and Russell (1992) *Quaternary* Proceedings, 2, figure 7, p. 142]

BOX 4.6: PREDICTING THE MAGNITUDE OF JÖKULHLAUPS

Attempts have been made to predict the peak flow magnitude of jökulhlaups, both for theoretical purposes and for use in design of bridges or structures that may have to withstand jökulhlaup flows. Developing theoretical relationships to predict discharge from ice-dammed lakes has proved difficult, not least because the exact mechanisms of drainage are poorly understood in many cases. However, Clague and Mathews (1973) developed an empirical equation

based on data from the jökulhlaups for which the peak discharge could be
estimated accurately. They obtain a regression relationship between peak
discharge and the volume of the ice-dammed lake.

$$Q_{max} = 75 \left(\frac{V_o}{10^6} \right)^{0.67}$$

where Q_{max} is peak discharge ($m^3\,s^{-1}$) and V_0 is the volume of the lake prior to
discharge (m^3).

This equation gives surprisingly good results and has been used widely to
predict jökulhlaup flood magnitudes.

Source: Clague, J.J. and Mathews, W.H. (1973). The magnitude of jökulhlaups. *Journal of*
    Glaciology, **12**, 501–4.

4.8 GLACIAL MELTWATER EROSION

Glacial meltwater erosion beneath ice sheets and glaciers may result from either
mechanical or chemical processes. The effectiveness of meltwater as an agent of
erosion depends on: (i) the susceptibility of the bedrock involved, in particular the
presence of structural weaknesses or its susceptibility to chemical attack; (ii) the
discharge regime, in particular the water velocity and the level of turbulent flow;
and (iii) the quantity of sediment in transport. Here we outline the main processes
of glacial meltwater erosion. The landforms created by the flow of meltwater are
described in Chapter 6.

4.8.1 Mechanical Erosion

Mechanical erosion occurs through two processes: (i) fluvial abrasion; and
(ii) fluvial cavitation.

Fluvial abrasion occurs by the transport of both suspended sediment and sediment in traction within the meltwater. This sediment abrades the walls of rock
channels and the bed of subglacial tunnels, striating and grooving the rock surface.
The rate of meltwater abrasion is controlled by the following factors.

1. **The properties of the sediment in transport.** Of particular importance is the
   hardness of the sediment relative to the bedrock surface over which the meltwater is flowing. The harder the sediment in transport is relative to the bedrock
   beneath the meltwater the more erosion is achieved. The concentration of debris
   within the meltwater is also important. In general rates of abrasion increase with
   increasing concentrations of debris over the range of concentration normal in
   most meltwater.

2. **Flow properties.** The rate of fluvial abrasion increases with the flow velocity. Similarly, the more turbulent the water flow the greater the rate of abrasion because sediment particles are brought into contact with the bed and channel walls more frequently than when the level of turbulence is low.

3. **Properties of the channel.** The roughness and orientation of facets within a channel as well as its planform all affect the rate of fluvial abrasion. Erosion is greatest where sediment-charged water impacts at a near normal angle. Consequently, obstructions within the channel, such as large boulders, will be rapidly abraded.

Fluvial abrasion may also occur without the transport of rock debris. Boulders that are too large or are wedged together may be vibrated by the passage of water. This vibration may cause abrasion or attrition of one boulder against the next or against the channel walls. Similarly boulders or stones trapped within enclosed hollows may achieve considerable amounts of abrasion as they are swirled around within the hollow.

Fluvial cavitation occurs wherever the meltwater velocity exceeds about $12\,\mathrm{m\,s^{-1}}$. It involves the creation of low-pressure areas within turbulent meltwater as it flows over a rough bedrock surface. These low-pressure areas form as the flow is accelerated around obstacles on the channel floor. If the pressure within the water drops sufficiently to allow the water to vaporise, bubbles of vapour (cavities) form. The cavitation bubbles grow and are moved along in the fluid until they reach a region of slightly higher local pressure where they will suddenly collapse. If cavity collapse occurs adjacent to a channel wall, localised but very high impact forces are produced against the rock. Repetition of these impact forces may lead to rock failure. In particular the shock waves are often forced into microscopic cracks within a rock or between mineral grains, causing them to loosen and allowing their removal. Once a surface has become pitted or fretted from the loss of grains or small rock fragments the shock waves tend to become concentrated or guided into the pits, thereby accelerating the process of erosion. Large bowl-shaped depressions may be produced in this way.

4.8.2 Chemical Erosion

Glacial meltwater can also erode bedrock by the processes of chemical solution. Soluble components of rock and rock debris are dissolved and removed in solution. This process is particularly important on carbonate-rich lithologies (e.g., limestone and chalk), but is not restricted to them. Chemical denudation beneath glaciers is often neglected as a process of glacial erosion, although, in recent years its importance has been increasingly recognised. Chemical denudation is particularly effective beneath glaciers, despite the low temperatures and therefore reaction rates, for three main reasons.

1. **High flushing rates.** Meltwater passes through the glacial system rapidly and is rarely stored subglacially for long; its residence time is therefore short, and this ensures that it does not have time to become chemically saturated.

2. **Availability of rock flour.** Turbulent meltwater is able to transport large quantities of freshly ground rock particles in suspension, which provide a very high surface area, or reaction surface, over which solution can occur.

3. **Enhanced solubility of carbon dioxide at low temperatures.** The solution of carbon dioxide by meltwater produces a weak acid. The solubility of carbon dioxide increases at low temperatures and consequently meltwater becomes more acidic and therefore more aggressive.

Chemical denudation is restricted to warm-based ice with abundant meltwater and may be particularly important in maritime areas where high rainfall adds significantly to the volume of water passing through the glacial system.

4.9 SUMMARY

Glacial meltwater is a very important component of the glacial system. It is the main ablation product of most ice sheets, it is intimately linked to glacier motion through sliding and subglacial sediment deformation, and it is responsible for removing debris from the ice–rock interface and carrying it beyond the confines of the glacier. The type of drainage network carrying glacial meltwater within an ice sheet or glacier is dependent on its thermal regime. In cold glaciers drainage is supraglacial, whereas in warm glaciers drainage it is supraglacial, englacial and subglacial. On hard substrates subglacial drainage occurs through either or both a linked system of cavities, or via a few major subglacial conduits or tunnels. On soft substrates subglacial drainage either occurs through the deforming sediment or via shallow channels cut in the surface of the sediment. The direction of englacial and subglacial drainage within a glacier follows the water pressure gradient, which is perpendicular to lines of equal water pressure potential. These are controlled primarily by the surface slope of the glacier and to a lesser extent by subglacial topography. Meltwater discharge varies diurnally and seasonally. High-magnitude, low-frequency events, known as jökulhlaups, may be superimposed on this discharge fluctuation. These catastrophic events are caused either by subglacial volcanic activity or by the drainage of ice-dammed lakes. Glacial meltwater erosion takes place through both mechanical and chemical erosion.

SUGGESTED READING

The hydrology of glaciers is outlined by Paterson (1998) and in review papers by Röthlisberger and Lang (1987), Hooke (1989) and Fountain and Walder (1998). Hock (2005) reviews the processes of glacier melt. Hubbard and Glasser (2005) explain the techniques used to investigate glacier hydrology. Hubbard and Nienow (1998) review alpine subglacial hydrology, Hodgkins (1997) reviews high-Arctic glacier hydrology, and Fountain *et al.* (2005) explain how meltwater travels internally through glaciers. Estimating the water pressure potential gradient within glaciers is covered by Shreve (1972) and its relationship to subglacial landforms is discussed in Shreve (1985a,b), Sugden *et al.* (1991) and Syverson *et al.* (1994). Sharp *et al.* (1993) used dye tracing to reconstruct the subglacial hydrology of a small glacier in the European Alps. Holmlund (1998) discusses the internal geometry and evolution of moulins, and Gulley and Benn (2007) describe how englacial conduits develop from structures within the ice. The reorganisation of groundwater flow in aquifers by ice sheets is dealt with in Boulton *et al.* (1993).

Jansson *et al.* (2003) review water storage in glaciers. The patterns of water and sediment discharge from glaciers have been described in papers by Willis *et al.* (1996), Singh *et al.* (2004) and Swift *et al.* (2005). Roberts (2005), Russell *et al.* (2006) and Evatt *et al.* (2006) review the characteristics of jökulhlaups and subglacial floods. Russell (1989) describes the drainage of two ice-dammed lakes in Greenland, and Tweed and Russell (1999) discuss the drainage mechanisms of ice-dammed lakes more generally. Clayton and Knox (2008) describe the drainage of a large proglacial lake that formed along the southern margin of the Laurentide Ice Sheet.

The relationship between subglacial water storage and glacier motion is explained at the valley glacier scale by Bartholomaus *et al.* (2008) and at the ice sheet scale by Bell (2008). Boulton *et al.* (2007) illustrate the movement of subglacial meltwater in groundwater channels, and Boulton *et al.* (2001) describe the role of this meltwater in subglacial sediment deformation. Meltwater quality is reviewed by Brown (2002), and Collins (1979) provides a case study of how meltwater quality can be used to investigate subglacial hydrology on alpine glaciers. Hubbard *et al.* (1995) explain how boreholes can be used to infer glacier drainage properties. A wider description of the role of glacial meltwater in mechanical and chemical erosion is given by Drewry (1986) and Fairchild *et al.* (1994) and Sharp *et al.* (1995) provide valuable case studies. Roberts *et al.* (2002) and Cook *et al.* (2006) examine the role of supercooled meltwater in the glacier system. Finally, Arnold and Sharp (1992) show how subglacial hydrology was important in determining the behaviour of Pleistocene ice sheets.

Arnold, N. and Sharp, M. (1992) Influence of glacier hydrology on the dynamics of a large Quaternary ice sheet. *Journal of Quaternary Science*, **7**, 109–24.

Bartholomaus, T.C., Anderson, R.S. and Anderson, S.P. (2008) Response of glacier basal motion to transient water storage. *Nature Geoscience*, **1**, 33–7.

Bell, R.E. (2008) The role of subglacial water in ice-sheet mass balance. *Nature Geoscience*, **1**, 297–304.

Boulton, G.S., Slott, T., Blessing, K., *et al.* (1993) Deep circulation of groundwater in over-pressured subglacial aquifers and its geological consequences. *Quaternary Science Reviews*, **12**, 739–45.

Boulton, G.S., Dobbie, K.E. and Zatsepin, S. (2001) Sediment deformation beneath glaciers and its coupling to the subglacial hydraulic system. *Quaternary International*, **86**, 3–28.

Boulton, G.S., Lunn, R., Vidstrand, P. and Zatsepin, S. (2007) Subglacial drainage by groundwater–channel coupling, and the origin of esker systems: Part 1—glaciological observations. *Quaternary Science Reviews*, **26**, 1067–90.

Brown, G.H. (2002) Glacier meltwater hydrochemistry. *Applied Geochemistry*, **17**, 855–83.

Clayton, J.A. and Knox, J.C. (2008) Catastrophic flooding from Glacial Lake Wisconsin. *Geomorphology*, **93**, 384–97.

Collins, D.N. (1979) Quantitative determination of the subglacial hydrology of two alpine glaciers. *Journal of Glaciology*, **23**, 347–62.

Cook, S.J., Waller, R.I. and Knight, P.G. (2006) Glaciohydraulic supercooling: the process and its significance. *Progress in Physical Geography*, **30**, 577–88.

Drewry, D. (1986) *Glacial Geologic Processes*, Arnold, London.

Elliston, G.R. (1973) *Symposium on the hydrology of glaciers*, International Association of Scientific Hydrology, **45**, figure 1, p. 80.

Evatt, G.W., Fowler, A.C., Clark, C.D. and Hulton, N.R.J.H. (2006) Subglacial floods beneath ice sheets. *Philosophical Transactions of the Royal Society*, **364**, 1769–94.

Fairchild, I.J., Brady, L., Sharp, M. and Tison, J. (1994) Hydrochemistry of carbonate terrains in alpine glacial settings. *Earth Surface Processes and Landforms*, **19**, 33–54.

Fountain, A.G. and Walder, J.S. (1998) Water flow through temperate glaciers. *Reviews of Geophysics*, **36**, 299–328.

Fountain, A.G., Schlicting, R.B., Jacobel, R.W. and Jansson, P. (2005) Fractures as main pathways of water flow in temperate glaciers. *Nature*, **433**, 618–21.

Gulley, J. and Benn, D.I. (2007) Structural control of englacial drainage systems in Himalayan debris-covered glaciers. *Journal of Glaciology*, **53**, 399–412.

Hock, R. (2005) Glacier melt: a review of processes and their modelling. *Progress in Physical Geography*, **29**, 362–91.

Hodgkins, R. (1997) Glacier hydrology in Svalbard, Norwegian High Arctic. *Quaternary Science Reviews*, **16**, 957–73.

Holmlund, P. (1988) The internal geometry and evolution of moulins, Storglaciären, Sweden. *Journal of Glaciology*, **34**, 242–8.

Hooke (1989) Englacial and subglacial hydrology: a qualitative review. *Arctic and Alpine Research*, **21**, 221–33.

Hubbard, B.P. and Glasser, N.F. (2005) *Field Techniques in Glaciology and Glacial Geomorphology*, John Wiley & Sons, Chichester.

Hubbard, B.P. and Nienow, P. (1998) Alpine subglacial hydrology. *Quaternary Science Reviews*, **16**, 939–55.

Hubbard, B.P., Sharp, M.J., Willis, I.C., *et al.* (1995) Borehole water-level variations and the structure of the subglacial drainage system of Haut Glacier d'Arolla, Valais, Switzerland. *Journal of Glaciology*, **41**, 572–83.

Jansson, P., Hock, R. and Schneider, T. (2003) The concept of glacier storage: a review. *Journal of Hydrology*, **282**, 116–29.

Paterson, W.S.B. (1998) *The Physics of Glaciers*, 3rd edn, Pergamon, Oxford.

Roberts, M.J. (2005) Jökulhlaups: a reassessment of floodwater flow through glaciers. *Reviews of Geophysics*, **43**, RG10002.

Roberts, M.J., Tweed, F.S., Russell, A.J. *et al.* (2002) Glaciohydraulic supercooling in Iceland. *Geology*, **30**, 439–42.

Röthlisberger, H. and Lang, H. (1987) Glacial hydrology, in *Glacio-Fluvial Sediment Transfer*, (eds A.M. Gurnell and M.J. Clark), John Wiley & Sons, Chichester, pp. 207–84.

Russell, A.J. (1989) A comparison of two recent jökulhlaups from an ice-dammed lake, Søndre Strømfjord, west Greenland. *Journal of Glaciology*, **35**, 157–62.

Russell, A.J., Roberts, M.J., Fay, H., *et al.* (2006) Icelandic jökulhlaup impacts: implications for ice-sheet hydrology, sediment transfer and geomorphology. *Geomorphology*, **75**, 33–64.

Sharp, M., Richards, K., Willis, I., *et al.* (1993) Geometry, bed topography and drainage system structure of the Haut Glacier D'Arolla, Switzerland. *Earth Surface Processes and Landforms*, **18**, 557–71.

Sharp, M., Tranter, M., Brown, G.B. and Skidmore, M. (1995) Rates of chemical denudation and CO_2 drawdown in a glacier-covered alpine catchment. *Geology*, **23**, 61–4.

Shreve, R.L. (1972) Movement of water in glaciers. *Journal of Glaciology*, **11**, 205–14.

Shreve, R.L. (1985a) Late Wisconsin ice-surface profile calculated from esker paths and types, Katahdin esker system, Maine. *Quaternary Research*, **23**, 27–37.

Shreve, R.L. (1985b) Esker characteristics in terms of glacier physics, Katahdin esker system, Maine. *Geological Society of America Bulletin*, **96**, 639–46.

Singh, P., Haritashya, U.K., Ramasastri, K.S. and Kumar, N. (2004) Diurnal variations in discharge and suspended sediment concentration, including runoff-delaying characteristics, of the Gangotri glacier in the Garhwal Himalayas. *Hydrological Processes*, **19**, 1445–57.

Sugden, D.E., Denton, D.H. and Marchant, D.R. (1991) Subglacial meltwater channel system and ice sheet over riding of the Asgard range, Antarctica. *Geografiska Annaler*, **73A**, 109–21.

Swift, D.A., Nienow, P., Hoey, T.B. and Mair, D.W.F. (2005) Seasonal evolution of runoff from Haut Glacier d'Arolla, Switzerland and implications for glacial geomorphic processes. *Journal of Hydrology*, **309**, 133–48.

Syverson, K.M., Gaffield, S.J. and Mickelson, D.M. (1994) Comparison of esker morphology and sedimentology with former ice-surface topography, Burroughs glacier, Alaska. *Geological Society of America Bulletin*, **106**, 1130–42.

Tweed, F.S. and Russell, A.J. (1999) Controls on the formation and sudden drainage of glacier-impounded lakes: implications for jökulhlaup characteristics. *Progress in Physical Geography*, **23**, 79–110.

Willis, I.C., Richards, K.S. and Sharp, M.J. (1996) Links between proglacial stream suspended sediment dynamics, glacier hydrology and glacier motion at Midtdalsbreen, Norway. *Hydrological Processes*, **10**, 629–48.

5: The Processes of Glacial Erosion

The processes by which a glacier detaches, picks up and transports rock and sediment as it moves across the Earth's surface are known as *glacial erosion*. Glaciers and ice sheets are agents of net erosion because as they flow towards their margins, they remove and transport debris within them. This reduces the elevation of the land under the ice sheet by erosion and increases it by deposition near the ice-sheet margin, where material is released. Material that has been entrained into the basal layers of the glacier is known as *basal debris*.

The processes and mechanics of glacial erosion are relatively poorly understood. This is because these processes occur deep beneath a glacier and are difficult to observe or measure. Direct observations of glacial erosional processes come either from subglacial tunnels or in ice-marginal cavities. Consequently, research on this topic has advanced along theoretical lines or by inferring process from the analysis of the landforms produced by glacial erosion in deglaciated areas. Two principal mechanisms of subglacial erosion are usually recognised: (i) *glacial abrasion*, the gradual wearing down of the bed by the passage of ice armed with debris; and (ii) *glacial plucking* or *quarrying*, the removal of blocks of rock or sediment from the glacier bed. A third process, *glacial meltwater erosion*, caused by the flow of meltwater beneath a glacier, was described in Chapter 4.

5.1 GLACIAL ABRASION

Glacial abrasion is the process by which rock particles transported at the base of a glacier are moved across a bedrock surface, scratching and wearing it away. This process is often likened to the action of sandpaper as it is scraped over a block of wood. Glacial abrasion creates striations, rock-scoured features and rock polish (see Section 6.1.1). It has been argued that glacial abrasion rates vary with the size of the material in traction. Large clasts (≥ 0.01 m) embedded in the glacier erode the bed primarily by scratching, creating *striations* and *grooves*, whereas finer material (≤ 0.01 m), especially the silt fraction, causes *polishing* of bedrock surfaces. There is compelling theoretical evidence that polishing of bedrock surfaces by fine material is quantitatively more important than scratching by larger clasts in the erosion of bedrock surfaces, although this remains to be confirmed by field observations. Experiments and observations from beneath modern glaciers (Box 5.1) suggest that three main variables control the ability of a glacier to abrade its bed. They are: (i) the basal contact pressure between the rock in the sole of the glacier and the bed; (ii) the rate of basal sliding; and (iii) the concentration and supply of rock fragments within the sole of the glacier. Each of these variables is examined below.

BOX 5.1: DIRECT OBSERVATIONS AND MEASUREMENTS OF GLACIAL ABRASION

The direct observation of abrasion in action is extremely difficult because it involves digging tunnels through a glacier to access basal cavities. Some of the first observations are those of Boulton (1974) who described the movement of a basalt fragment over a large basalt roche moutonnée 20 m below the surface of Brei∂amerkurjökull in southeast Iceland. The basalt fragment was removed from the base of the glacier and the surface that had been in contact with the bed was inspected. The fragment had been in contact with the bed at three points and between the points of contact crushed debris had been ploughed up in front. Striations produced by the fragment could be traced for 3 m. The largest striation was seen to deepen rapidly to 3 mm but then to gradually shallow to 1 mm. Boulton (1974) related this decrease in depth to the build up of crushed debris, which spreads the load at the interface over a wider area. The build up of a layer of crushed debris was thought to result in a change in the nature of motion from a jerky 'stick-slip' motion, to a relatively uniform sliding movement – the stick-slip motion produces a carpet of debris over which the particle subsequently slides. When this carpet is exhausted by comminution, the clast will again come into contact with the bed, thereby recutting the striation. This may explain the disappearance and reappearance of striae. These observations suggest that there are two abrasive processes: (i) the cutting of striae; and (ii) polishing of the bed by fine debris that is ploughed up when a striation is cut. Boulton (1974) went on to measure the rate of abrasion by cementing rock and metal plates to bedrock surfaces adjacent to basal cavities beneath Brei∂amerkurjökull in Iceland and the Glacier d' Argentiere in the French Alps. These plates became quickly covered by basal ice and were later recovered for inspection (see Table below). Boulton's results have now been supplemented by measurements beneath 200 m of ice at Engabreen, Norway (Box 5.2). Here, Cohen *et al.* (2005) measured the friction between the debris in basal ice and a smooth tablet of rock inserted under the glacier. The ice contained 10% debris by volume and exerted local shear traction of up to 500 kPa. Calculations show that the shear traction due to the friction between the debris and the bed is around 100 kPa at Engabreen. These authors concluded that the friction between debris in basal ice and the bed is much higher than previously assumed and is sufficient to have a retarding effect on rates of glacier sliding.

Locality	Average abrasion rate (mm yr^{-1})		Ice thickness (m)	Ice velocity (m yr^{-1})
	Marble plate	Basalt plate		
Breiðamurkerjökull 1	3	1	40	9.6
Breiðamurkerjökull 2	3.4	0.9	15	19.5
Breiðamurkerjökull 3	3.75		32	15.4
Glacier d' Argentiere	36		100	250

Sources: Boulton, G.S. (1974) Processes and patterns of glacial erosion, in *Glacial Geomorphology* (ed. D.R. Coates), *Proceedings of the Fifth Annual Geomorphology Symposia*, Binghampton, Allen & Unwin, London, pp. 41–87. Cohen, D., Iverson, N.R., Hooyer, T.S., *et al.* (2005) Debrisbed friction of hard-bedded glaciers. *Journal of Geophysical Research – Earth Surface*, **110**, F02007.

5.1.1 Basal Contact Pressure

This is probably the most important variable that determines the rate of glacial abrasion. Using the analogy of a block of wood being sanded with a piece of sandpaper, it follows that the harder you press down on the surface of the wood the faster it is worn away. In a glacier this is the contact pressure between the clast in basal transport and the glacier bed beneath it: the greater the pressure the more abrasion will occur. There are, however, two alternative views on what controls this contact pressure and the movement of basal clasts. The first view was developed by Geoffrey Boulton and the second was suggested later by Bernard Hallet.

1. **The Boulton model.** This model assumes that the contact pressure between a particle in contact with the glacier bed is related to the *effective normal pressure*. As we saw in Section 4.6, this is a function of: (i) *normal pressure*, given by the weight of the overlying ice; and (ii) the *basal water pressure*, which acts in opposition to the weight of the overlying ice by buoying up the glacier, similar to the action of a hydraulic jack. Effective normal pressure therefore will be high when: (i) the ice is thick; and (ii) basal water pressure is low. This last point is of some importance because in a cold-based glacier, where there is little or no meltwater present at the glacier bed, effective normal pressure will be much higher than for a warm-based glacier of similar thickness. Similarly bedrock lithology beneath a warm-based glacier may also be important.

Effective normal pressures will be much higher on porous rocks because this will reduce the basal water pressure (see Section 4.6). Where the bed is not horizontal, for example where there is a bedrock obstacle, effective normal pressure is modified by an amount equal to the pressure of the ice flowing against the obstacle (Figure 4.6: see Section 4.6).

In the Boulton model effective normal pressure controls the rate of abrasion. As effective normal pressure increases, abrasion will also increase as the clast in the base of the ice is being pushed harder into the bed. However, as effective normal pressure increases, the friction between the clast and the bed is also increased and this friction will ultimately begin to slow the movement of the particle and the basal ice that holds it will begin to flow around the clast. When this occurs, abrasion will start to decrease despite the fact that effective normal pressure is still increasing. Consequently, if all other variables are held constant (e.g., sliding velocity) then abrasion will first increase with effective normal pressure and then decrease until the friction between the clast and the bed is such that it will stop moving and will lodge.

2. **The Hallet model.** This model assumes that the contact pressure between a clast in basal transport is independent of the effective normal pressure. This theory is based on the premise that clasts are completely surrounded by ice and can be considered to be essentially floating within it. This occurs because ice will deform around a clast by creep due to the weight of the ice above it and therefore basal clasts are effectively surrounded by ice at all times. In this case the contact pressure between a clast and the glacier bed is a function of the rate at which ice flows towards the bed, forcing the clast into contact with the bed. This depends on: (i) the rate of basal melting; and (ii) the presence of extending glacier flow. In this model, abrasion is independent of variations in effective normal pressure and is primarily a function of basal melting. As we saw in Section 3.4 basal melting is favoured by: (i) rapid ice flow, which generates large amounts of frictional heat; (ii) thick ice; (iii) high ice surface temperatures; and (iv) the advection of warm ice towards the glacier bed.

We shall return to these two different views of basal contact pressure below because two very different models of glacial abrasion have been developed around them.

5.1.2 Basal Sliding

The rate of basal sliding controls the rate at which basal debris is physically dragged across the surface below, and consequently the greater the rate of basal sliding the greater the amount of abrasion. Using the sandpaper analogy, the faster you move the sandpaper back and forth then the faster the wood is worn away. Basal thermal regime is important (see Section 3.4), because sliding is not widespread beneath cold-based glaciers and these glaciers have less ability to abrade their beds (see Boxes 6.8 and 6.9).

5.1.3 The Concentration and Supply of Rock Fragments

The concentration of debris within basal ice also controls the rate of abrasion. Ice on its own cannot cause significant abrasion – it needs debris within it to do this. The rate of abrasion, however, is not increased simply by increasing the concentration of basal debris. In fact it has been suggested that abrasion is most effective where basal debris is relatively sparse. This is because basal debris increases the frictional drag between the ice and its bed and therefore reduces the sliding velocity. Glaciers with relatively clean basal ice are able to slide faster than those with large amounts of basal debris. There is a certain threshold of debris concentration above which the abrasion rate declines with increasing debris content, because of its adverse effect on the rate of basal sliding (Figure 5.1).

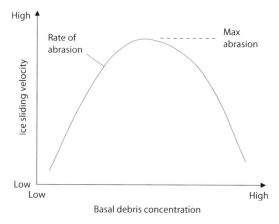

Figure 5.1   Schematic representation of the relationship between basal sliding velocity and the concentration of debris at the base of a glacier.

The type and shape of basal debris are also important. Some rocks are more durable than others, and a glacier armed with basal debris derived from a hard or resistant lithology will be more effective than one armed with a relatively soft lithology. The most effective combination occurs where a glacier armed with debris entrained from a hard substrate flows over a relatively soft lithology. If the debris is softer than the substrate little abrasion would occur, because erosion would preferentially reduce the size of the basal clasts first. The shape of the basal debris is also important because sharp fragments are able to make deeper incisions into the underlying bedrock than those with blunter or more rounded points or edges. Laboratory observations have shown that clasts in contact with the bed frequently rotate or flip, which helps to improve their life span as erosive tools beneath the glacier.

A continued supply of basal debris is also important because basal debris is quickly worn down and crushed. For abrasion to be effective, basal debris must therefore be continually replaced. This may occur either by: (i) the entrainment of fresh glacial debris at the glacier bed; or (ii) by basal melting, which progressively lowers debris down through a glacier towards the bed (see Section 7.2).

5.1.4 Abrasion Models

We have already seen that there are two alternative views concerning the nature and controls on the contact pressure between a basal clast and the bedrock beneath a glacier. Both Boulton and Hallet have developed numerical models with which to predict the patterns and amounts of glacial abrasion. The two models are very different.

Boulton's abrasion model assumes that the contact pressure on a rock particle at the base of a glacier is a function of the normal effective pressure. As a consequence his model predicts that abrasion will be controlled by: (i) the effective normal pressure; and (ii) the ice velocity. Effective normal pressure is controlled by ice thickness and basal water pressure (see Section 4.6). The relationship between abrasion and these two variables within Boulton's model is illustrated in Figure 5.2. This graph shows that for a given ice velocity abrasion increases to a peak as effective normal pressure increases, and then falls rapidly to zero as the friction between debris and bed becomes sufficient to retard the movement of the

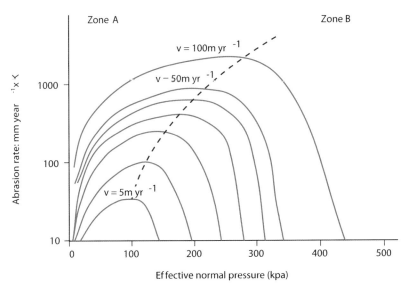

Figure 5.2   Graphic representation of Boulton's abrasion model. The graph shows theoretical abrasion rates plotted against effective normal pressure for different ice velocities. In Zone A, abrasion rates increase with increasing pressure whereas in Zone B abrasion rates decline with increasing pressure. Zone C, located to the right of the higher x-axis intercept for any one ice velocity, is an area of no abrasion and basal debris is deposited as lodgement till. [Modified from: Boulton (1974) in *Glacial Geomorphology* (ed. D.R. Coates), George Allen and Unwin, figure 7, p. 52]

particle. At effective normal pressures above a critical level no abrasion occurs, but instead debris hitherto transported is deposited. Erosion and deposition appear therefore to be two parts of a continuum.

Boulton has used this model to predict the evolution of bedrock bumps by glacial abrasion (Figure 5.3). In Section 4.6 we saw how effective normal pressure varied across an obstacle (Figure 4.6). Given this pattern of variation Boulton used his abrasion model to predict how the shape of a two-dimensional obstacle would change with erosion. He assumed that the bump had a sinusoidal shape, that the ice velocity over the bump was 50 m per year and that the pressure fluctuation over the bump was 130 kpa (Figure 4.6). Given these values he charted the evolution of two bedrock bumps under a glacier: one that had an effective normal pressure of 70 kpa and one which experienced 240 kpa. The two patterns of evolution are quite different and are shown in Figure 5.3. The bump with an

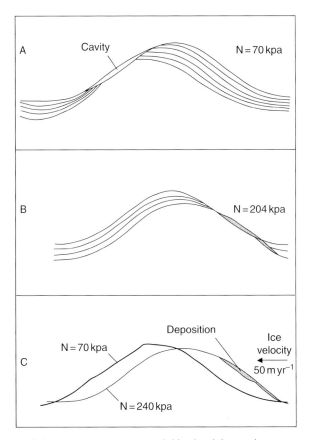

Figure 5.3 Patterns of abrasion across a sinusoidal bedrock bump for a constant ice velocity of 50 m per year for two different values of normal pressure using Boulton's abrasion model. [Modified from: Boulton (1974) in *Glacial Geomorphology* (ed. D.R. Coates), George Allen and Unwin, figure 9, p. 56]

effective normal pressure of only 70 kpa evolved into a stoss-and lee-form (see Section 6.2.2). The normal pressures all fall within the zone of rising abrasion and pressure (Zone A: Figure 5.2). The rate of abrasion was therefore highest on the up-glacier flank, low on the crest of the bump, zero on its lee flank where a cavity forms and high at the foot of the lee flank. A roche moutonnée shape is therefore produced with time (Figure 5.3A). The morphology of the bump under 240 kpa of effective normal pressure evolves very differently. Here the normal pressures fall within zones of falling or zero abrasion with increasing pressure (Zones B and C: Figure 5.2). Lodgement occurs on the up-glacier flank, slight abrasion occurs on the crest of the bump, maximum abrasion occurs on the down-glacier flank and slight abrasion at the foot of this flank. The effect is to produce a form that migrates in an up-glacier direction, with a steep up-glacier flank similar to a crag and tail (Figure 5.3B). In summary, therefore, the key implications of Boulton's abrasion model are:

1. Variations in ice thickness control abrasion and lodgement via effective normal pressure.
2. Variations in basal water pressure, controlled by such factors as bed permeability (geology) control abrasion via effective normal pressure.
3. Ice velocity controls the rate of abrasion.
4. Abrasion and lodgement form part of a continuum.

Hallet used an alternative approach to Boulton in formulating his abrasion model. Basal rock particles are envisaged as essentially floating hydrostatically in the ice and are therefore independent of effective normal pressure. In this model the rate of basal melting and ice velocity are the key controls on abrasion. As we will see in the next chapter Hallet's model has been used widely in numerical models designed to study the evolution of large glacial landforms. The main implications of Hallet's model are:

1. Abrasion is highest where basal melting is greatest.
2. Abrasion is independent of effective normal pressure and therefore of basal water pressure, although not glacier thickness because this controls the rate of basal melting.
3. Lodgement and abrasion are independent processes.

These two models contain very different predictions and at first sight these two theories seem to conflict. It is, however, possible that each model represents different but equally valid subglacial conditions. Boulton's model and predictions may apply where the basal ice is particularly dirty and debris-rich and therefore likely to behave as a solid slab. The rigid nature of this slab prevents the ice deforming around each clast. In contrast, Hallet's model and predictions may be more appropriate in areas where basal debris is sparse and the ice is consequently less rigid.

5.2 GLACIAL QUARRYING

Glacial plucking or *glacial quarrying* is the means by which a glacier removes larger chunks and fragments from its bed. It involves two separate processes: (i) the fracturing or crushing of bedrock beneath the glacier; and (ii) the entrainment of this fractured or crushed rock.

5.2.1 Fracturing of Bedrock Beneath a Glacier

The propagation of fractures within bedrock is essential for glacial quarrying. Fractures may pre-date the advance of a glacier into an area and simply reflect the geological structure of the area, or may be generated by periglacial freeze–thaw weathering in advance of the glacier. Glaciers may also create their own fractures in bedrock as they flow across it (Box 5.2).

BOX 5.2: DIRECT OBSERVATIONS OF SUBGLACIAL QUARRYING

Like glacial abrasion, the direct observation of glacial quarrying is extremely difficult because it involves digging tunnels through a glacier to access basal cavities. However, a 2 km-long artificial tunnel under the glacier Engabreen in Norway, originally built as part of a hydroelectric scheme, provides direct access to the bed of this temperate glacier some 200 m beneath the glacier surface. A number of important experiments have been conducted here on subglacial processes, including studies of glacial abrasion, glacial quarrying, water pressure variations, ice rheology and debris concentrations. To study glacial quarrying, Cohen *et al.* (2006) installed a granite step, 120 mm high with a crack in its stoss surface, at the bed of Engabreen and used acoustic emission sensors to monitor crack growth events in the step as the ice slid over it. This is illustrated in the photograph below which shows a rock step that was subjected to water pressure fluctuations in its lee over an 8 day period beneath ~ 215 m of ice at Engabreen. A 31 mm deep, 2 mm wide crack was cut in the rock's upper surface. The crack propagated downward over the 8 days of the experiment, such that upon removal of the step from the bed its lee surface was quarried, as shown. Acoustic emission source locations, shown in the lower figure, show the slow crack propagation with time, which was stimulated by water pressure changes and cavity growth in the step's lee. They also measured vertical stresses, water pressure and cavity height in the lee of the step. By artificially pumping water to the lee of the step they observed that adding water initially caused the lee-side cavity to open. The cavity then closed after pumping was stopped and water pressure decreased. During cavity closure, acoustic emissions from the base of the crack increased dramatically. With repeated pump tests this crack grew over time until the lee surface of the rock step was quarried. These experiments confirm that fluctuating water pressure in cavities is important in glacial quarrying because it greatly aids the development of cracks in the bedrock.

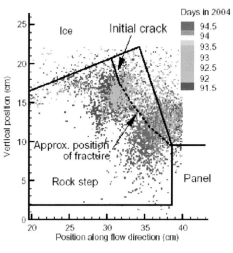

Source: Cohen, D., Hooyer, T.S., Iverson, N.R., *et al.* (2006) Role of transient water pressure in quarrying: A subglacial experiment using acoustic emissions. *Journal of Geophysical Research – Earth Surface*, **111** (F3), F03006. [Modified from: Cohen *et al.* (2006) *Journal of Geophysical Research – Earth Surface*, **111** figure 8 and 9, F03006]

Pressure release as glacial erosion proceeds may generate fractures and joints parallel to the erosional surface. As rock surfaces are unloaded they may expand and fracture. The removal of significant quantities of overlying rock by glacial erosion may cause unloading and the development of such fractures. This process has been used to explain the presence of large *sheet joints* parallel to eroded surfaces such as valley sides (Figure 5.4). These joints are produced by pressure release due to the unloading effect of glacial erosion. Glacial erosion may therefore generate bedrock fracture and thereby accelerate rates of erosion.

As a glacier moves over an irregular bedrock surface, complex patterns of basal ice pressure are generated (Figure 4.6). This pattern of pressure differences is transmitted to the underlying bedrock beneath, causing stress fields to be set up within the bedrock. These stress patterns are often more pronounced if a cavity exists in the lee of a bedrock obstacle. These stress fields may be sufficient to cause

Figure 5.4 Sheet joints formed by pressure release in granite on the side of a glacial valley in the Cordillera Blanca, Peru. Note how the surface of each sheet is parallel to the side of the valley.
[Photograph: N.F. Glasser]

the bedrock to fracture, although theoretical calculations have suggested that this will occur only where pre-existing joints and weaknesses exist (Box 5.3). A rock mass with such weaknesses is referred to as a *discontinuous rock mass*. All bedrock

---

BOX 5.3: CRUSHING OF BEDROCK OBSTACLES BENEATH A GLACIER

Morland and Morris (1977) used a mathematical model to study the potential for bedrock crushing by a glacier. Their aim was to see whether the stress field produced in bedrock by an overriding glacier was sufficient to cause failure in the bedrock. They calculated the likelihood of bedrock failure for different object shapes and different bedrock lithologies. The results of one experiment designed to predict the region of a bedrock hump where bedrock failure is most likely is shown below. The maximum stress generated by the glacier moving over this bump is located deep within the rock mass on the down-stream flank of the bump. The value of this failure stress is less than the coherent strength of the bedrock itself, which led Morland and Morris (1977) to conclude that failure will not occur if the rock is coherent. In this situation the profile of a bedrock hummock, such as that illustrated below, will remain stable unless bedrock joints or other internal weaknesses are present within the rock. Morland and Morris (1977) concluded that these weaknesses there-fore must be present to allow bedrock failure to occur and to facilitate the development of a typical roche moutonnée profile.

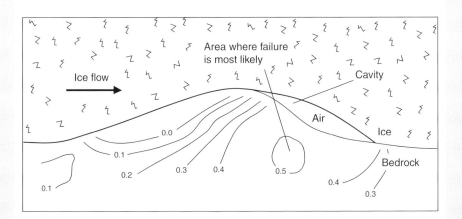

Source: Morland, L.W. and Morris, E.M. (1977) Stress in an elastic bedrock hump due to glacier flow. *Journal of Glaciology*, **18**, 67–75. [Modified from: Morland and Morris (1977) *Journal of Glaciology*, **18**, figure 7, p. 74]

contains joints, bedding planes and other lines of weakness, which may be exploited and expanded by the stress field generated by the flow of ice over them. The importance of these weaknesses is reflected in the fact that the morphology of many glacial erosional landforms is controlled by the pattern of discontinuities, joints and bedding planes within the parent rock mass (Figure 5.5). In general, the faster the rate of ice flow the more pronounced are the variations in basal ice stress and therefore the stress field generated within the underlying bedrock. The pattern of effective normal pressure over obstacles may fluctuate with time, causing the stress fields within the bedrock obstacle to vary. Dramatic changes or repeated changes in the stress fields may be particularly important in propagating fractures along lines of weakness within the rock mass.

Figure 5.5   (A) Ice margin of the Greenland Ice Sheet showing blocks of bedrock removed by glacial quarrying. (B) Blocks of bedrock removed from a rock face by glacial quarrying, Norway. Former ice flow was from right to left. In both cases the size of the quarried blocks is controlled by the joint spacing in the bedrock. [Photographs: N.F. Glasser]

Temporal fluctuations in the pattern of ice pressure may also cause variation in the temperature of basal ice and in some cases may generate small cold-based patches within an otherwise warm-based glacier. This process is known as the *heat-pump effect*. At its simplest this process involves the melting of ice in areas of high basal ice pressure, for example on the upstream side of an obstacle (see Figure 4.6). The high pressure reduces the freezing point, allowing melting to occur. Melting of ice consumes thermal energy, latent heat, and will cause the ice mass to cool. Some or all of the meltwater generated will move under the glacier to areas of lower basal ice, pressure where it refreezes. If the basal ice pressure then falls over the original obstacle, refreezing of the available meltwater will occur around this obstacle. On freezing, latent heat is given off and will warm the basal ice. However, because some of the meltwater has now been lost the temperature of the basal ice cannot regain its former level and a cold patch will form. The same amount of water would need to refreeze as was melted to return the basal ice to its original temperature, but since meltwater has flowed away this cannot occur. Consequently temporal pressure variations beneath an ice sheet, associated for example with diurnal fluctuations in ice velocity or meltwater discharge, may generate cold patches on the glacier bed. Beneath these cold patches ice will be frozen to the bedrock, creating *sticky spots* at the bed in these places. Lumps of rock beneath cold patches may therefore be entrained by freezing to the glacier as the ice flows forward (see Section 5.2.2).

Fluctuations in basal water pressure may also help to propagate bedrock fractures beneath a glacier (Figure 5.6). Basal water pressure influences fracturing in two ways: (i) it affects the distribution and magnitude of the stress fields set up by ice in bedrock surfaces; and (ii) its presence within fractures and microscopic cracks is important to the process of fracture propagation. As we saw in Section 4.6, basal water pressure helps determine the presence or absence of basal cavities beneath a glacier. The presence or absence of these basal cavities has an important influence of the distribution of stresses imposed on a bedrock obstacle by the glacier. Changes in the basal water pressure within lee-side cavities causes them to vary in size and may cause cavity closure, and thereby alter the stress field within the bedrock

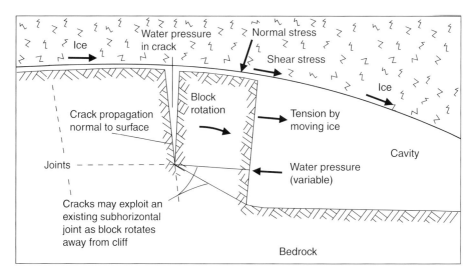

Figure 5.6   Schematic diagram of the processes involved in glacial quarrying. Fluctuations in subglacial water pressure are important in the development of bedrock fractures, and blocks are rotated out of the rock face by the passage of ice.

obstacle. For example, the sudden drainage of meltwater from within a cavity may cause it to close even if the ice velocity remains unchanged, due to increase in the effective normal pressure caused by the fall in the water pressure (see Section 4.6). Rapid or repeated changes of this sort may help to widen or propagate fractures within the bedrock mass (Figure 5.6). As a cavity forms due to an increase in water pressure this hydrostatic pressure can effectively lift up the base of the glacier. If rock fragments are frozen to the glacier bed due to the heat-pump effect during this process they will be lifted up and moved forward as the base of the glacier rises (Figure 5.7). This process is known as the *hydraulic-jack effect*.

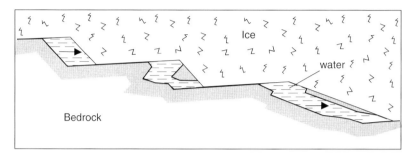

Figure 5.7   The role of the hydraulic-jack and heat-pump effects in glacial quarrying. Fragments of bedrock are frozen to the glacier bed by the heat-pump effect and then lifted from it as increasing water pressure opens basal cavities.

Where fluctuations of basal water pressure are combined with the heat-pump effect a cycle of erosion may occur. As basal water pressure falls the resulting increase in effective normal pressure over an obstacle may cause a period of fracture propagation. As basal water pressure rises basal cavities may open and rock may be entrained by the hydraulic-jack effect and the heat-pump process. This would imply that glacial quarrying will be most effective under a glacier where there are regular fluctuations in basal water pressure.

5.2.2 Rock and Debris Entrainment

Debris entrainment includes the processes by which bedrock or sediment is detached from the glacier bed and incorporated into the basal ice. The evacuation and entrainment of a rock fragment from the glacier bed is governed by the balance between the tractive force exerted on it by the overriding ice and the frictional forces which act to hold the rock in place. High basal water pressures may help to reduce the frictional forces holding debris in place. Entrainment can occur in the following ways.

1. By the heat-pump effect, causing local patches of basal ice to freeze to the bed and therefore detaching debris as the ice flows forward.
2. The drag between ice and bedrock may be sufficient to detach very loose particles, particularly if they become surrounded by ice.

3. Loose debris collected within basal cavities may become surrounded by ice and simply swept away if the cavity closes.

4. Freezing-on of material may occur in the lee of obstacles as meltwater generated on their upstream faces refreezes in the low-pressure zone in their lee to form regulation ice. In warm-based glaciers the debris layer produced by the freezing of regulation ice is usually thin because debris is also released by melting on the upstream side of obstacles. However, if freezing-on dominates over melting, perhaps at boundary between warm and cold ice, where there is a constant flux of meltwater freezing onto the glacier, a large thickness of debris-rich regulation ice may develop (Figure 5.8).

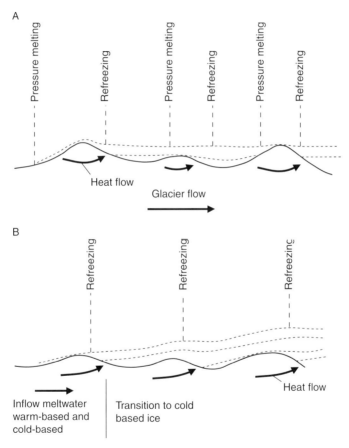

Figure 5.8   Debris incorporated into regulation ice in the lee of bedrock bumps. (A) In a warm-based glacier regulation ice and debris layers tend to be destroyed by pressure melting associated with other bumps and the debris layer will be thin. (B) In a zone of transition from warm-based to cold-based ice, meltwater and debris can be frozen into the basal ice layer and significant thicknesses of debris can form. [Modified from: Boulton (1972) in: *Polar Geomorphology* (eds R.J. Price and D.E. Sugden), Institute of British Geographers, Special Publication 4, figure 5, p. 101]

5. Through refreezing associated with *glaciohydraulic supercooling*, a process where subglacial meltwater flowing in a distributed drainage system can 'supercool' (i.e. exist as liquid water at a temperature below its freezing point) as it moves up an adverse bed slope of an *overdeepening*. At contemporary glaciers (e.g., in Alaska and Iceland) this process can create thick zones of debris-laden basal ice in areas where the bed slope exceeds the ice-surface slope by more than 1.2–1.7 times. It has also been suggested that this process was widespread beneath the former Laurentide and Scandinavian Ice Sheets, for example in areas with overdeepenings close to the former ice margin, but this has yet to be substantiated.

6. Beneath glaciers with a mixed thermal regime, fluctuation of the thermal boundary between warm and cold ice may lead to the freezing-on of large rafts of sediment or rock. For example, if an area of previously warm basal ice was to turn cold, perhaps during deglaciation, large rafts of previously saturated sediment and bedrock may become frozen to the glacier bed and entrained as it flows forward.

7. Debris may also be incorporated into the ice along thrust planes. In glaciers with a mixed thermal regime in which there is a cold ice margin and warmer interior, compressive flow is common because the cold ice moves less quickly than the warm ice within the glacier interior. This compression may lead to the development of thrusts within both the ice and underlying sediment along which debris may be incorporated (Figures 7.11 and 7.12: see Section 7.3).

5.3 ESTIMATING RATES OF GLACIAL EROSION

In the previous sections we examined the principal processes of glacial erosion. The efficiency of these processes coupled with the length of time over which they operate determines the amount of glacial erosion that can be achieved by a glacier. In this section we examine the methods available with which to calculate the depths and rates of glacial erosion. There are four ways in which estimates of the rates of glacial erosion can be obtained.

1. **Direct observation beneath modern glaciers.** Direct observations and experiments at the base of present-day glaciers can be conducted by drilling or tunnelling to the base of a glacier. This process is facilitated if access can be obtained via such tunnels to subglacial cavities. In these cavities and tunnels, measurements of glacial abrasion can be obtained by periodically measuring either the microtopography of bedrock surfaces or with reference to artificial plates fixed to the rock surface beneath the glacier. Rates of glacial erosion observed beneath present-day glaciers in Iceland have given rates for glacial abrasion of between 0.5 and 1 mm per year (Box 5.1). No quantitative observations have yet been made of the process of glacial quarrying beneath modern glaciers. This process operates periodically and is therefore difficult to monitor over a short time period.

2. **Geomorphological reconstructions of erosion in glaciated terrains.** Geomorphological observations and the distribution of preglacial deposits can

be used to reconstruct the form of the preglacial landscape. This reconstructed surface is then compared with the present glacial surface to give a measure of the depth of glacial erosion (Box 5.4). This is a useful tool in areas where preglacial remnants are abundant, but is of limited use if the preglacial landscape has been largely removed. The more intense the glacial erosion the more difficult it becomes to quantify the depth of erosion. In areas of multiple glaciation the erosion estimate reflects net erosion and also includes any fluvial erosion that may have taken place during interglacial periods. It is also important to note that these methods rely on the extrapolation between preglacial fragments, and is therefore open to error. As a consequence this method may be used only to provide order-of-magnitude estimates of glacial erosion.

BOX 5.4: RATES OF GLACIAL EROSION FROM PREGLACIAL RECONSTRUCTIONS

The geomorphological reconstruction of preglacial landscapes provides a tool with which glacial geologists can estimate the depth and rate of glacial erosion. In the Sognefjord area of western Norway there is strong evidence that much of the landscape is preglacial in origin and has simply been dissected by a system of glacial troughs (fjords). Nesje *et al.* (1992) attempted to reconstruct the depths of erosion in this area by reconstructing the preglacial landscape. They assumed that the area was once dominated by a preglacial dendritic valley system, fragments of which survive on the inter-fluves between the glacial fjords. These fragments were used to reconstruct the preglacial landscape, as shown below. From this reconstruction of the preglacial surface, they were then able to subtract the present-day topography to determine the depth of Quaternary erosion, assumed to have been primarily glacial. In the study the total volume of rock removed equalled 7610 km^3. This corresponds to an average erosion depth of 610 m over the area. Although during the Quaternary ice age the area has been glaciated several times it is possible to estimate the total length of time that the area was glaciated from the sum of the duration of each period of glaciation. A figure of 600 000 years is obtained from these estimates. This corresponds to an average erosion rate of 1 mm per year.

Source: Nesje, A., Dahl, S.O., Valen, V. and Ovstedal, J. (1992) Quaternary erosion in the Sognefjord drainage basin, western Norway. *Geomorphology*, **5**, 511–20.

---

3. **Estimates of glacial erosion from cosmogenic radionuclide measurements on bedrock surfaces.** This is a relatively new technique which relies on using cosmogenic nuclides in bedrock to establish whether there are remnant nuclide concentrations in samples taken from landforms that represent relict areas and landforms that represent effective glacial erosion (see Box 6.10).

4. **Sediment volume calculations.** A variety of methods have been used to relate sediment output from a glacial catchment to the rate of erosion within it. This method has been used to estimate erosion for small valley glaciers by measuring either the total sediment output from a glacial catchment, meltwater sediment discharge and ice-marginal accumulation, or simply just one component such as the volume of suspended sediment in meltwater. If the sediment output from the glaciated catchment and the total catchment area are known then the amount of erosion necessary to produce the sediment output can be calculated (Box 5.5). This method cannot be used for large ice sheets where catchments are difficult to define. However, a similar strategy has been developed to calculate the erosion beneath former ice sheets from the sediment deposited in offshore

---

BOX 5.5: GLACIAL EROSION BENEATH MODERN GLACIERS INFERRED FROM SEDIMENT OUTPUT

There are two commonly used methods by which calculations of the rate of glacial erosion can be obtained from the sediment output of modern glaciers. The first method is to measure sediment concentrations in proglacial streams leading from the glacier over a given period of time. Riihimaki *et al.* (2005) documented sediment evacuation rates for three seasons at Bench Glacier, a small temperate alpine glacier in the Chugach Range of Alaska, to see if sediment evacuation rates track glacial erosion rates over annual time scales. They found that much of the sediment emerges in a major pulse that closely follows the termination of enhanced sliding each year, suggesting that the conduits and well-connected cavities that allowed removal of englacial water to terminate sliding also promoted efficient transport of sediment. They concluded that the glacier lowers its bed by 1–2 mm per year while sliding only 1–2 m per year. The second method is to examine the rate of glacial sedimentation directly. Small *et al.* (1984) surveyed the moraine embankments at the margins of the Glacier de Tsidjiore Nouve in Switzerland. This enabled them to calculate the growth of the moraines over a 5 year period. From this they calculated the volume of sediment required to produce the changes in moraine volume and estimated that marginal sedimentation was of the order of 8727–11 230 tonnes per year. To this they added the volume of sediment moved each year as suspended load in the proglacial streams in front of the glacier. Averaged over the entire catchment area of the glacier this is equivalent to an erosion rate of 1.55–2.29 mm per year. Both these estimates of glacial erosion are consistent with other estimates obtained for Alpine glaciers.

Source: Riihimaki, C.A., MacGregor, K.R., Anderson, R.S., *et al.* (2005) Sediment evacuation and glacial erosion rates at a small alpine glacier. *Journal of Geophysical Research – Earth Surface*, **110** (F3), F03003. Small, R.J., Beecroft, R.I. and Stirling, D.M. (1984) Rates of deposition on lateral moraine embankments, Glacier de Tsidjiore Nouve, Valais, Switzerland. *Journal of Glaciology*, **30**, 275–81.

3. Warm-based zones of an ice sheet experience high levels of basal melting and therefore greater levels of glacial abrasion than areas with low rates of basal melting.

4. Areas subject to fluctuations in water pressure are likely to experience greater amounts of glacial quarrying than areas with little or no fluctuation.

5. It has been suggested that an ice sheet which grows in a previously glaciated area with an ice divide located in the same place as in the past may achieve less erosion than an ice sheet growing for the first time within an area. Once an efficient network of glacial troughs and discharge routes has been established there is less resistance to ice discharge and therefore less erosion. It follows therefore that in areas glaciated several times most of the glacial erosion may be completed during the first phase of glaciation.

It is these second-order variables that give rise to the complex and local variations we find within landscapes of glacial erosion. It is also important to note that there are a number of important feedbacks between glacier thermal regime, sediment supply, meltwater routing and the rates of sediment transport by meltwater that also determine the efficiency of glacial erosion (Box 5.7).

BOX 5.7: STABILISING FEEDBACKS ON GLACIAL EROSION

Glaciologists have historically lacked knowledge about what governs glacial erosion on a basin-wide scale, hindering their attempts to model the processes of glacial erosion. By outlining a theoretical framework for 'graded glaciers', Alley *et al.* (2003) presented a set of rules such as those that govern flow rates, erosion and deposition in rivers. These authors considered the possibility that there are self-limiting or stabilising feedbacks in glacial erosion because glacier surface and bed slope are linked to glacier thermal regime, sediment supply, meltwater routing and the rates of sediment transport by meltwater. These feedbacks control the balance between erosion and deposition beneath glaciers. Crucial to their analysis is the role played by overdeepenings near the glacier terminus because in these areas subglacial meltwater channels become plugged by freezing water, sediment transport capacity drops dramatically and the overdeepenings begin to fill with sediment. As the overdeepenings begin to fill up with sediment, the glacier surface slope reduces and the system may switch back to meltwater flow and erosion. This cycle would act as a stabilising feedback. Alley *et al.* hypothesised that the long profiles of the beds of highly erosive glaciers tend towards steady-state angles opposed to and slightly more than 50% steeper than the overlying ice–air surface slopes, although this remains to be tested quantitatively beneath both contemporary and former glaciers.

Source: Alley, R.B., Lawson, D.E., Larson, G.J., *et al.* (2003) Stabilizing feedbacks in glacier-bed erosion. *Nature*, **424**, 758–60.

5.5 SUMMARY

Glacial erosion is the removal of fragments of rock or sediment from a glacier bed. It may occur through three processes: (i) glacial abrasion; (ii) glacial quarrying; and (iii) by the action of glacial meltwater. Glacial abrasion involves the scouring action of debris dragged over a bedrock surface by a glacier and is controlled primarily by effective normal pressure and ice velocity. Glacial quarrying involves the fracturing and entrainment of bedrock beneath a glacier. The nature of the glacial erosion undertaken by a particular glacier is dependent on its basal thermal regime because this is the fundamental control on the supply of meltwater to the base of the glacier and therefore dictates the presence or absence of basal sliding. Unless the ice is warm-based there will be no basal sliding and the processes of glacial erosion cannot act. As basal thermal regime changes over time, rates of glacial erosion may also vary through time. Estimates of rates of glacial erosion are derived from direct subglacial observations, geomorphological reconstructions and from the volume of sediment removed from a glaciated catchment. These studies confirm that fast-flowing, warm-based glaciers achieve significantly more erosion than slow-flowing, cold-based glaciers.

SUGGESTED READING

Subglacial observations of glacial erosional processes in basal cavities and tunnels are recorded in several papers: Boulton (1974), Vivian (1980), Anderson *et al.* (1982), Rea and Whalley (1994) and Cohen *et al.* (2005, 2006). Boulton (1974, 1979) discusses the fundamental principles of glacial erosion and introduces a model of glacial abrasion. Alternative abrasion models are outlined by Hallet (1979, 1981) and Hindmarsh (1996a). The supply of basal debris to the base of a glacier and its role in glacial abrasion is considered by Röthlisberger (1968) and Hindmarsh (1996b). The heat-pump effect is introduced and described by Robin (1976). This work is expanded by Röthlisberger and Iken (1981), and Iverson (1991) concentrates on the importance of fluctuations in subglacial water pressures and their role in glacial quarrying. The fracturing of bedrock beneath glaciers is treated numerically by Morland and Morris (1977) and by Addison (1981), and Atkinson (1982, 1984) provides a good introduction to the role of bedrock conditions in fracture propagation within bedrock itself. Rea (1994) provides a good overview of rock fracture and glacial quarrying, and Cuffey *et al.* (2000) discuss entrainment at cold glacier beds. The factors that control rock resistance to erosion are outlined by Augustinus (1991). A readable discussion of the links between the glacier system, mass balance and glacial erosion is provided by Andrews (1972).

Harbor and Warburton (1993) discuss the debate over rates of erosion in glacial and non-glacial areas. Good examples of specific studies attempting reconstructions of depths and rates of glacial erosion are those of Sugden (1978), Nesje *et al.* (1992), Hall and Sugden (1987) and Hallet *et al.* (1996). Links between patterns of glacial erosion and basal thermal regime are explored in papers by Sugden (1974,

| p-forms and s-forms | Smooth-walled sculpted depressions and channels cut in bedrock. Encompasses a variety of morphological expressions including sichelwannen, hairpin erosion marks, potholes, bowls, channels and grooves. | Landforms carrying striae indicate warm-based ice carrying a basal debris load and high clast–bed contact pressures (>1 MPa) inferred from intimate ice–bedrock contact. Landforms where striae are absent indicate presence of abundant basal meltwater, possibly concentrated by catastrophic discharge. Low effective normal pressures inferred for these landforms. |

[Modified from: Glasser, N.F. and Bennett, M.R. (2004). *Progress in Physical Geography*, **28**, 43–75.]

Figure 6.2 Examples of microscale landforms of glacial erosion. (A) Smoothed and abraded bedrock outcrop with two sets of cross-cutting striations in front of a glacier in Svalbard. (B) Striations on the bedrock outcrop in front of a glacier in the Khumbu Himalaya. Note that the bedrock is also fractured along joints and that it is also stained by subglacial precipitates. (C) Striations on a facetted boulder in front of the Tasman Glacier, New Zealand. (D) Micro crag and tail on a bedrock surface in Canada. Former ice flow was right to left. (E) Long, continuous striations on a bedrock surface in western Ireland. (F) Crescentic fractures on a bedrock surface in western Ireland. [Photographs: N.F. Glasser]

Figure 6.2   continued

Figure 6.2   continued

Figure 6.2 continued

of the bedrock surface on which they are located, in particular the presence of stoss and lee forms (see Section 6.2.2); (ii) the presence of micro crags and tails; (iii) the presence of friction cracks; or (iv) relationship to other larger landforms.

Striations are usually not more than a few millimetres in depth, but may be over several metres long. Their continuity is broken by small gaps or breaks where contact between the bed and clast was temporarily broken during their formation (Box 6.1). This may occur due to the formation of small subglacial cavities or alternatively where a clast rides up over a cushion of debris. The depth and

BOX 6.1: THE MORPHOLOGY OF GLACIAL STRIAE AND INFERENCES FOR GLACIAL ABRASION

Iverson (1991) used glacial striae, recently exposed on carbonate bedrock adjacent to Saskatchewan Glacier, Canada, to make inferences about the mechanics of glacial abrasion. Iverson measured the width, depth and length of individual striae, as well as making observations about a range of morphological criteria. The shapes of striae indicate that abrading fragments commonly rotate. Iverson defined three types of striae. *Type 1 striae* become progressively wider and deeper down-glacier until they end abruptly, often as deep steep-walled gouges. They are inferred to form as a striating clast ploughs forward and downward, before either the striator point breaks off the clast or the torque on the clast is sufficiently large so that it rotates out of the groove. *Type 2 striae* start and terminate as faint, thin traces. They steadily broaden and deepen until they reach a maximum width and depth near their centre point. They are probably formed by sharp striator points that are rotating as they slide. The point initially has a large ploughing angle, causing progressive incision of the striation. Deeper ploughing causes more rapid clast rotation as the torque on the clast increases. Rotation, together with comminution of the clast point, reduces the ploughing angle so that there is a steady reduction in striation depth. Consequently Type 2 striae indicate clasts that slow down until the maximum striation depth is reached and then steadily accelerate, until at the striation terminus, the clast has the same velocity as the ice. *Type 3 striae* begin abruptly as deep gouges and then become progressively narrower and shallower down-glacier. They are inferred to form where a striator point contacts and indents the bed. Clast rotation with little displacement along the bed produces a low ploughing angle, so that a gradual reduction in indentation depth occurs as sliding proceeds. Overall, the conclusion is that clasts with steep leading edges will abrade progressively deeper into the bed with sliding, whereas those with more gently inclined leading edges will climb out of their grooves. The paper shows how, therefore, glacial striae can be used to make inferences about former subglacial processes.

Source: Iverson, N.R. (1991) Morphology of glacial striae – implications for abrasion of glacier beds and fault surfaces. *Geological Society of America Bulletin*, **103**, 1308–16.

continuity of striations is a balance between the effective normal pressure, which keeps the base of the glacier in contact with its bed, and changes in basal water pressure, which allow small cavities to form (see Section 4.6). Striations formed by different ice-flow directions may be superimposed in a cross-cut pattern (*cross-cut striations*; Figure 6.2A). This occurs when the ice-flow direction changes, either due to a readvance of ice over a deglaciated area or due to changes in ice-flow direction within a glacier. Cross-cut striations record the fact that the second ice flow was unable to erode all the evidence of the earlier flow, either because of a lower efficiency of glacial abrasion or due to insufficient time. The occurrence of cross-cut striations can be used, therefore, to make inferences about former glacier dynamics (Box 6.2).

BOX 6.2: USING LANDFORMS OF GLACIAL EROSION TO RECONSTRUCT GLACIER DYNAMICS

Landforms of glacial erosion are often used by glacial geologists to infer basic glaciological parameters such as ice-movement direction and change over time, but are seldom used for anything more complex. Sharp *et al.* (1989) demonstrated for the first time how glacial erosional landforms might be used to reconstruct parameters that affect the operation of basal processes and glacier dynamics. Their field study area is Snowdon, North Wales: an area that has been subjected to multiple glaciations of different duration and intensity. During the height of the last glacial maximum the Snowdon area was overrun by an ice sheet with a divide located in mid-Wales. At the close of the last glacial cycle, however, small cirque glaciers existed on Snowdon during a period known as the Younger Dyras (or Loch Lomond) Stadial. These two glacial episodes are marked by cross-cut striations. Detailed mapping of the glacial erosional landforms present, in particular of the size of striations and position of former lee-side cavities on bedrock surfaces, allowed Sharp *et al.* (1989) to suggest that the processes of glacial erosion were different during the two glacial events. Erosion beneath the ice sheet was dominated by lee-side fracturing of bedrock obstacles, by surface fracturing that created friction cracks, and by widespread abrasion. In contrast, erosion during the second phase, by the cirque glaciers, was confined to glacial abrasion and there is little evidence for lee-side cavities. Many of the surface features eroded on the bedrock surface by the ice sheet were not removed by the later cirque glacier, suggesting that relatively little erosion took place during this final episode of glaciation. From this, Sharp *et al.* (1989) infer that the cirque glacier had low sliding velocities which prevented cavity formation. On the basis of these inferences and mass balance estimates, Sharp *et al.* (1989) were able to calculate the dynamics of these former ice bodies. The cirque glacier was shown to have a low sliding velocity, around 10 m per year, high contact pressure at the glacier base, low basal water pressures and therefore few lee-side cavities. In contrast the earlier ice sheet was shown to have a much higher sliding velocity (>35 m per year), lower basal contact pressures, higher basal water pressures and therefore widespread cavity formation. This illustrates the detailed inferences that can be made using simple observations of glacial erosional landforms combined with numerical estimates.

Source: Sharp, M., Dowdeswell, J.A. and Gemmell, J.C. (1989) Reconstructing past glacier dynamics and erosion from glacial geomorphic evidence: Snowdon, North Wales. *Journal of Quaternary Science*, **4**, 115–30.

Since striations are produced by glacial abrasion they indicate that: (i) the ice contained basal debris; (ii) the ice was warm-based and moved by basal sliding; (iii) there were moderate levels of normal effective pressure; and (iv) there was transport of rock debris towards the bed by basal melting. If any of these prerequisites for glacial abrasion ceases to exist, then the formation of striations will also cease. In this situation, the striated surface will become fossilised and the last ice-flow direction will be left imprinted on the bedrock surface. There are three situations where this may occur.

1. Where deglaciation occurs and the bedrock becomes re-exposed as the glacier margin retreats.

2. Where the basal thermal regime beneath a glacier changes from warm-based to cold-based. Once basal sliding stops, the basal ice becomes frozen to the bedrock beneath. In this situation new striations cannot form and the existing striations are preserved.
3. Where a layer of basal till is deposited immediately on top of a striated bedrock surface. In this scenario, the last ice-flow direction suggested by the striations will correspond approximately to the age of the overlying till. Striations therefore may be placed in a stratigraphical framework if the relative ages of the till units within an area can be established.

It is worth noting at this point that, under certain conditions, striations can also form under cold-based glaciers. Examples have been described from the Allan Hills area of Antarctica (see Box 6.9). Here the striations are shorter, less continuous and more irregular than those formed under warm-based conditions.

Striations can be used to reconstruct local patterns of ice flow, but their application to large-scale ice-flow reconstruction is more problematic. This is because basal ice conditions, and especially basal thermal regimes, change markedly over both time and space and consequently the pattern of striations beneath a glacier may be asynchronous (i.e. composed of a variety of different ages). Striations located in close proximity on the bed of a former glacier therefore may date from different time periods or relate to different ice flows with radically different flow directions. Furthermore, the preservation of striations on the bed of a former ice sheet depends upon the basal boundary conditions during deglaciation: for example, whether deglaciation occurs beneath cold-based ice (preservation of striations), or beneath warm-based ice (new striations forming constantly during deglaciation).

Striations formed during warm-based deglaciation will change direction as the orientation of the ice margin changes during recession. The youngest striations on a

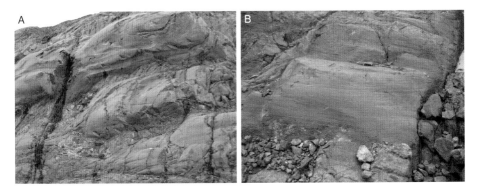

Figure 6.3 Ice-smoothed valley walls next to the San Rafael Glacier in Chile. (A) Smoothed bedrock showing evidence of both polishing by glacial abrasion and the development of fractures by glacial quarrying. Former ice flow left to right. (B) Close-up of the valley walls showing smoothing by subglacial meltwater and glacial abrasion (note the striations). Former ice flow left to right. [Photographs: N.F. Glasser]

bedrock outcrop will be orientated perpendicular to the ice margin, whereas older striations are related to more distant ice-marginal positions. Consequently, the further a striation is from the current ice margin, the more difficult it becomes to relate its formation to a particular ice-flow event. This situation is complicated further if cold-based deglaciation occurs, because this inhibits the formation of new striations and leads to the preservation of older striations that can be connected to the current ice margin. These complications mean that the interpretation of striation patterns over large areas, such as an entire ice-sheet bed, is a complicated task.

6.1.2 Micro Crag and Tails

Micro crag and tails are small tails of rock, which are preserved from glacial abrasion in the lee of resistant grains or mineral crystals on the surface of a rock. For example, in the slate rock of North Wales the presence of occasional pyrite crystals forms a point of resistance in an otherwise homogeneous rock. In the lee of these pyrite crystals small tails of rock are preserved. In many cases the pyrite weathers out on deglaciation (Figure 6.2D). Micro crag and tails are important because they provide clear evidence of both the orientation and direction of ice flow.

6.1.3 Friction Cracks

Friction cracks are a family of small cracks, gouges, chatter marks and indentations created in bedrock as larger boulders or clasts beneath a glacier are forced into contact with the bed. They vary in form from crescentic shaped gouges in which small chips of bedrock have been removed to fracture lines or cracks. Three main types of feature can be recognised: (i) *crescentic fractures*, which are a series of small cracks often forming a distinct line that are usually convex up-ice (Figure 6.2F); (ii) *crescentic gouges*, which occur where crescentic chips of rock have been removed and are normally concave up-glacier; and (iii) *chatter marks*, which are a series of irregular fractures. These features are not always consistently orientated in the direction of ice flow. For example, crescentic gouges are occasionally convex up-ice, when they are referred to as *reverse crescentic gouges*. In general there is much morphological diversity to these features and a wide variety of different forms have been recorded. They tend to form preferentially on crystalline or homogeneous bedrock lithologies.

Friction cracks differ from striations because they are not produced by the continuous contact between a clast and the glacier bed. Instead, they are formed by intermittent ice–bed contact. Local variations in effective normal pressure and bedrock topography are sufficient to make a clast 'bounce' or roll over a bedrock surface, creating small gouges or cracks when the clast comes periodically into contact with the bed. Friction cracks provide evidence of high effective normal pressures because considerable contact force between the clast and the bedrock is required to cause bedrock fracturing.

6.1.4 P-Forms and Micro Channel Networks

Smooth sinuous depressions and large grooves sculpted in bedrock are given the collective term *plastically moulded forms* or *p-forms*. The most commonly encountered types of p-forms are *sichelwannen, potholes* or *bowls* and *channels* (Figure 6.4). Sichelwannen are sickle-shaped bedrock depressions, usually occurring with an open end that points in the direction of ice flow. These open ends may be extended in the direction of ice flow as shallow runnels. Individual features are normally around 1 m in length but may occasionally exceed 10 m. Where they are particularly elongate in morphology they are referred to as *hairpin erosional marks*. These features are found at a wide range of different sizes from a few millimetres to several metres. *Potholes* are more rounded and deeper depressions which often occur in conjunction with sichelwannen. They may be up to several metres in diameter and depth. Sinuous or linear *channels* cut into bedrock, such as *Nye channels*, are also common (Figure 6.4A). These channels are usually less than a metre in width and depth. All these features may occur either in isolation or in close association and may be found with striations and other features of glacial abrasion. Striations are sometimes found superimposed on p-forms.

Figure 6.4 Examples of p-forms and s-forms. (A) Deep Nye channel in front of Glacier de Ferpècle in Switzerland (former ice and water flow away from the camera). [Photograph: B.P. Hubbard] (B) Large grooves formed by a combination of glacial abrasion and meltwater erosion on a bedrock surface in Patagonia. [Photographs: N.F. Glasser]

The origin of p-forms is a source of debate. There are three main hypotheses: (i) formation by glacial abrasion; (ii) formation due to abrasion by a till slurry; and (iii) formation by meltwater. The presence of glacial striations on some p-forms has led many to argue for a mechanism involving glacial abrasion. The organisation of basal debris into distinct lines or streams at the base of a glacier (see Figure 7.8) would tend to concentrate glacial abrasion in certain areas, allowing the ice to sculpt grooves or channels. The alternative mechanisms involve either meltwater processes or a hyperconcentrated flow of till and water (a till slurry). For example, the formation of hairpin erosional marks and sichelwannen can be explained by flow separation around a small obstacle on a bedrock surface (Figure 6.5). The size of the obstacle controls the size of the erosional mark produced. They may form around single crystals, grains or nodules that protrude up through a bedrock surface, or alternatively around much larger bedrock knobs.

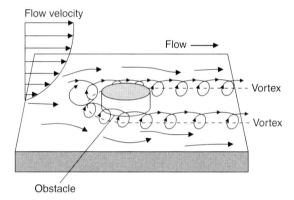

Figure 6.5   Horseshoe vortices formed as meltwater is diverted around an obstacle on the glacier bed. [Modified from: Shaw (1994) *Sedimentary Geology*, **91**, figure 8, p.276]

6.2 MESOSCALE FEATURES OF GLACIAL EROSION

Mesoscale features of glacial erosion are those between 1 m and 1 km in size and comprise a family of landforms which includes: (i) streamlined bedrock features; (ii) stoss and lee forms; (iii) rock grooves and basins; and (iv) meltwater channels (Table 6.2).

6.2.1 Streamlined Bedrock Features

At a mesoscale the most common effect of glacial erosion is to streamline bedrock protrusions to produce positive and upstanding landforms. These streamlined landforms, *whalebacks*, are also referred to by a wide variety of different terms

Table 6.2   Mesoscale landforms of glacial erosion and their significance for the reconstruction of former ice masses.

Landform	Morphology	Glaciological significance
Streamlined bedrock features ('whalebacks')	Streamlined bedrock eminences with abraded surfaces on all sides.	Warm-based ice carrying a basal debris load. High effective normal pressures (>1 MPa) inferred from intimate ice–bedrock contact and cavity suppression. Thick ice. Low sliding velocity with little available basal meltwater.
Stoss and lee forms ('roches moutonnées')	Upstanding bedrock eminence with both abraded and quarried faces. Detailed morphology controlled by preglacial weathering characteristics and patterns of bedrock jointing.	Warm-based ice carrying a basal debris load. Low effective normal pressures (0.1–1 MPa) inferred from the presence of basal cavities. Quarried faces indicate abundant basal meltwater with regular fluctuations in basal water pressure. Rapid sliding velocity. Some evidence that roches moutonnées form under thin ice, for example possibly during ice-sheet build-up and decay. May indicate direction and orientation of ice flow.
Rock grooves, bedrock megagrooves and rock basins	Smooth-walled sculpted depressions and channels cut in bedrock.	Warm-based ice carrying a basal debris load. Quarried landforms indicate low effective normal pressures (0.1–1 MPa) inferred from the presence of basal cavities with abundant basal meltwater and regular fluctuations in basal water pressure. Quarried landforms indicate rapid sliding velocity and thin ice. They may also indicate direction and orientation of ice flow. Landforms occurring in association with striae indicate high effective normal pressures (>1 MPa) inferred from intimate ice–bedrock contact and cavity suppression.
Subglacial meltwater channels	Steep sided channels cut into bedrock or till. Channel orientation may be discordant with the local topography. Channels may have an irregular convex-up long profile.	Warm-based ice carrying a basal debris load. Channel systems can be used to calculate former hydraulic potential gradient and therefore to infer the regional pattern of subglacial drainage, and to estimate ice-surface slope and ice thickness.

Ice-marginal meltwater channels	Either a complete channel cross-section or a channel floor and one wall wherever the other wall was formed by ice (half channel). Channels start and end abruptly. Often associated with other ice-marginal depositional landforms.	Calculations of palaeovelocity and palaeodischarge possible from measurements of channel shape, channel width and size of material transported by meltwater flow.
		Release of large quantities of supra-, en- or subglacial meltwater. Channels indicate the location of the former ice margin and patterns of ice recession. Gradient of channel long profile may indicate that of the ice margin. Calculations of palaeovelocity and palaeodischarge possible from measurements of channel shape, channel width and size of material transported.

[Modified from: Glasser, N.F. and Bennett, M.R. (2004). *Progress in Physical Geography*, **28**, 43–75.]

such as *rock drumlins*, *tadpole rocks* and *streamlined hills* (Figures 6.6 and 6.7). Whalebacks are bedrock knolls that have been smoothed and rounded on all sides by a glacier. Individual whalebacks may be slightly elongated in the direction of ice flow, although the structural attributes of the bedrock (e.g., joints, bedding planes and foliations) may dramatically affect the morphology of their overall form. Whalebacks tend to have low height to length ratios. They are relatively high (1–2 m) in comparison to their length (1.5–3 m). Striations and other small-scale features of glacial abrasion may be superimposed on any surface of a whaleback. Striations on whalebacks are often continuous along the entire length of the whaleback. From this it is possible to infer that there were no basal cavities around the whaleback during its formation and the ice was everywhere in contact with

Figure 6.6 Examples of streamlined bedrock features. (A) Whaleback in resistant granite, Norway (former ice flow left to right). (B) Roche moutonnée above Nant-y-Moch, Wales (former ice flow left to right). [Photographs: N.F. Glasser]

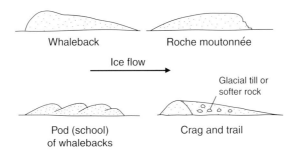

Figure 6.7   The four main types of streamlined glacial erosional landforms.

the landform so that landforms were produced primarily by glacial abrasion. High effective normal pressures therefore must be present on both the proximal and distal faces of the whaleback in order to suppress the formation of basal cavities.

6.2.2 Stoss and Lee Features

In contrast to whalebacks and other landforms dominated by glacial abrasion, stoss and lee features possess both abraded and plucked surfaces, and have therefore a pronounced asymmetry (Figures 6.6 and 6.7). They are defined as bedrock knolls or small hills with a gently abraded slope on the up-ice side (stoss) and a steeper, rougher plucked or quarried slope on the down-ice side (lee). The most common type of stoss and lee landform is a *roche moutonnée*. These landforms often occur in clusters or fields and may vary in size from several metres to tens or hundreds of metres. They form where high effective normal pressures occur on the stoss side of a bedrock hummock, but the pressure is sufficiently low on the down-ice side to allow a cavity to form (see Section 4.6). Consequently the up-ice side experiences glacial abrasion while the down-ice side is glacially plucked. The presence of a lee-side cavity is pre-requisite for the formation of a roche moutonnée and consequently they are restricted to areas where the ice flows fast enough and the effective normal pressure is sufficiently low to allow cavities to open. Roches moutonnées therefore form preferentially in areas of thin and fast flowing ice. As glacial quarrying is facilitated by regular and frequent fluctuations in basal water pressure, their formation is also facilitated by the presence of subglacial meltwater.

Once a lee-side cavity has opened and glacial plucking is initiated, the properties of the parent bedrock determine the detailed morphology of the resulting roche moutonnée (Figure 6.8). Bedrock jointing is particularly important because joint depth and spacing determine the size of the blocks that can be quarried from the lee of the original bedrock hummock. It is possible to predict the evolution of the plucked surface of a roche moutonnée (Figure 6.9). Block removal will begin at the furthest point down-ice in the cavity and as successive blocks are removed the

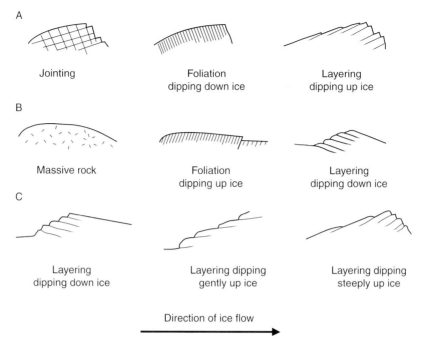

Figure 6.8 Schematic representation of the relationships between geological structure and roche moutonnée morphology. (A) Quarried lee-side slopes. (B) Abraded lee-side slopes. (C) Abraded and quarried lee-side slopes. [Modified from: Chorley *et al.* (1984) *Geomorphology*, Methuen, figure 17.18, p. 449]

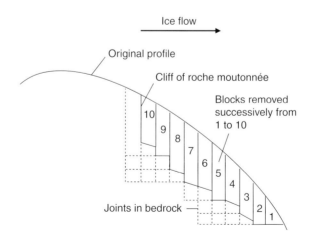

Figure 6.9 A theoretical model of the evolution of a roche moutonnée. Blocks are removed successively from number 1 to number 10. [Modified from: Sugden *et al.* (1992) *Geografiska Annaler*, **74A**, figure 4, p. 256]

quarried surface will migrate further up-ice. This results in a quarried lee-side face that resembles a staircase (Figure 6.9). The spacing of the horizontal and vertical joints within the bedrock will determine the dimensions of each step within the staircase. In bedrock that is not heavily jointed the glacier may create its own

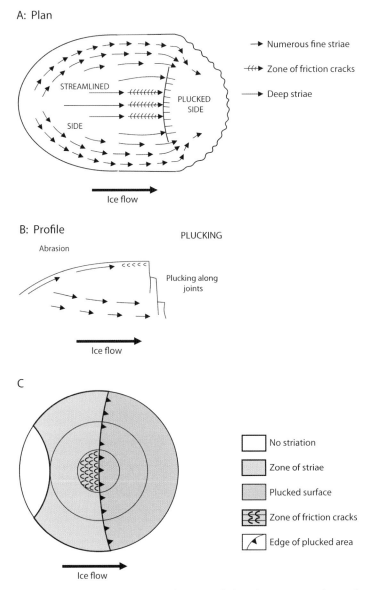

Figure 6.10   The distribution of microscale features of glacial erosion on the surface of roches moutonnées. (A and B) Plan and profile views of typical roches moutonnées. (C) A stereographic model of the distribution of microscale features of glacial erosion across the surface of a typical roche moutonnée. [Modified from: Chorley *et al.* (1984) *Geomorphology*, Methuen, figure 17.17, p. 448]

fractures due to basal pressure fluctuations. In this case the staircase formed by the removal of blocks may be more varied.

A variety of other microscale erosional landforms often occur on the surfaces of roches moutonnées (Figure 6.10). Although roches moutonnées are characterised by abraded stoss slopes and quarried lee sides the orientation of these two surfaces is not always a reliable indicator of ice movement direction because their detailed morphology is controlled in part by the pattern of joints or other weaknesses within the rock mass (Figure 6.8). This reduces their reliability as palaeo-ice-flow indicators.

6.2.3 Grooves and Rock Basins

Grooves can be formed by either glacial abrasion or by meltwater erosion. Grooves are similar in morphology to striations, except for their greater size and greater depth. They range from tens of metres to hundreds of metres in length and may be up to several metres wide and a metre deep. They are probably the product of glacial abrasion although the flow of meltwater can also be important. The location and orientation of individual grooves may also be influenced by the presence of structural weaknesses within the rock mass.

Rock basins are individual depressions carved in bedrock. They are often found in association with roches moutonnées and may fill with water on deglaciation to become lakes. Rock basins range in size from several metres to hundreds of metres in diameter. The development of these basins is controlled by the distribution of structural weaknesses within a rock mass, which can be exploited by glacial quarrying. The size and density of the basin is therefore usually a function of the spacing of joints, or other lines of weakness, within the rock mass.

The formation of rock basins beneath a glacier provides a good illustration of the role of *positive feedback* within landform development. The situation is as follows: as a glacier flows over an irregular bed it develops zones of compressional and extensional flow (see Figure 3.8). Zones of compressional flow will tend to transport material away from the bed, whereas extensional flow will cause ice to move towards the bed. Erosion by glacial quarrying will be limited in areas of extensional flow because bedrock blocks cannot be transported away from the bed easily, although abrasion may be facilitated by the increased contact pressure between basal clasts and the bed. Net erosion by quarrying is favoured in areas dominated by compressional flow because the eroded blocks can be transported away from the bed. Consequently, as a glacier flows over a slight depression it will first experience extensional flow on the up-ice side and then compressive flow on the down-ice side of the depression. The extensional flow component will increase basal pressure and abrasion of the basin floor whereas the compressional phase will facilitate block removal and plucking. The form of the basin, therefore, will be accentuated by erosion. This will in turn increase the degree of extensional and compressional flow experienced by the glacier as it flows over the basin, which will in turn accelerate its erosion. In this way the basin will grow as consequence of the positive feedback between basin erosion and compressional flow.

6.2.4 Meltwater Channels

The final group of mesoscale landforms of glacial erosion are meltwater channels (Figure 6.11). Meltwater channels can form in five environments (Table 6.3): (i) subglacially (beneath the ice); (ii) laterally (along the ice margin); (iii) in proglacial locations (in front of the ice), associated with the flow of water away from the glacier or out of ice-contact lakes; (iv) in supraglacial (ice surface) or englacial (within the ice) environments; and (v) from subglacial lake outbursts.

The orientation of *subglacial meltwater channels* is controlled by the hydraulic potential gradient within the glacier (see Section 4.6.2). This gradient is determined primarily by the glacier surface slope and secondarily by the topography beneath the glacier. As a consequence, subglacial meltwater channels may cut across or be orientated transverse to the surface contours and drainage patterns of the present-day topography. Subglacial meltwater is also able to flow uphill if driven by the hydraulic gradient within the glacier. Consequently, subglacial meltwater channels may not have constant gradients but may have an 'up and down', or 'humped' long profile.

In contrast, *ice-marginal* and *lateral meltwater channels* run parallel to the glacier front and are most commonly found in the ablation area of glaciers where rates of surface ablation and meltwater production are highest. The morphology of these channels is variable: for example they may stop suddenly as meltwater is diverted into the glacier via a crevasse or moulin. This process may be recorded by a *chute* or sudden right angle bend in the channel. Ice-marginal channels sometimes start in large bowl-shaped depressions, interpreted as plunge pools formed by water falling from the glacier into the channel in the same way as plunge pools form on rivers beneath waterfalls. Some lateral meltwater channels have a distinct

Figure 6.11   Photographs of typical meltwater channels: (A) Ice-marginal meltwater channels next to a small valley glacier in Svalbard. (B) Subglacial meltwater channel near Helsby in Cheshire, England. [Photographs: N.F. Glasser]

Table 6.3 Diagnostic criteria for classification of meltwater channels.

Subglacial	Lateral		Proglacial	Supraglacial/englacial
	Marginal	Submarginal		
Undulating long profile	Parallel with contemporary contours		Regular meander bends	Meander forms crescentic valley on face of hill
Descent downslope may be oblique	Form 'series' of channels parallel to each other		Occasional bifurcation	Low gradient
Descent downslope may form steep chutes	Approximately straight		Flows direct downslope	Sinuous
Complex systems – bifurcating and anastomosing	Perched on valley sides		Large dimensions- wide and deep	Approximately constant width
High sinuosity	May terminate in downslope chutes			
Abandoned loops	Absence of networks	May form networks		
Abrupt beginning and end	Gentle gradient	Steeper gradient (oblique downslope)		
Absence of alluvial fans	Parallel for long distance	Sudden changes in direction		
Cavity systems and potholes				
Ungraded consequences	May terminate abruptly			
Variety of size and form within the same connected system	May be found in isolation from all other glacial features			
Association with eskers				

[Modified from: Greenwood *et al.* (2007) *Journal of Quarternary Science*, **22**, Table 1, p. 641]

cross-sectional form, whereas others may simply have a cross-section that resembles a bench or step cut into the hill side. The longitudinal gradient of ice-marginal channels approximates that of the glacier surface at the time of their formation.

In proglacial settings, powerful streams draining the glacier may cut distinct channels or gorges as the water flows away from the ice margin. Aggradational landforms, including sandar and braided outwash channels, are also common. *Ice-dammed lakes* may drain over cols or ridges via *overspill channels*. These are usually channels cut into cols or in notches within ridges or hills surrounding an ice-dammed lake.

Finally, meltwater channel systems can also be cut by *subglacial lake outburst floods*, where large volumes of water drain catastrophically from beneath a glacier or ice sheet towards the ice margin. One of the most famous examples of this is the Labyrinth in the Dry Valleys of Antarctica (Box 6.3), but there are also examples in the northwestern North America associated with the drainage of Glacial Lake Missoula and examples associated with subglacial volcanic activity in Iceland (jökulhlaups).

BOX 6.3: GLACIAL MELTWATER CHANNELS IN THE LABYRINTH: EVIDENCE FOR HUGE SUBGLACIAL FLOODS FROM THE ANTARCTIC ICE SHEET

Perhaps the most famous meltwater channel system in the world is the Labyrinth, a >50 km long network of bedrock channels and scoured terrain in front of Wright Upper Glacier in Antarctica. Lewis *et al.* (2006) described in detail the geomorphology of the channels that make up the Labyrinth. It consists of a number of different levels. On the upper and intermediate levels are signs of glacial erosion in the form of striations and ice-moulding. On the lower level the channels and canyons are up to 600 m wide and 250 m deep, have longitudinal profiles with reverse gradients, and contain huge potholes (>35 m deep) at tributary junctions. Lewis *et al.* (2006) considered these characteristics to be consistent with incision from fast-flowing subglacial meltwater, possibly related to a former subglacial outburst flood associated with episodic drainage of subglacial lakes in East Antarctica. They estimated flood discharges on the order of 1.6 to $2.2 \times 10^6 \, \text{m}^3 \, \text{s}^{-1}$. Using $^{40}\text{Ar}/^{39}\text{Ar}$ analyses of volcanic tephra in the Labyrinth they then showed that the channels are relict and that the last major subglacial flood occurred sometime between 14.4 and 12.4 million years ago. They also speculated that the huge discharge of large volumes of subglacial meltwater to the Southern Ocean may have coincided with, and contributed to, oscillations in regional and/or global climate at this time.

Source: Lewis, A.R., Marchant, D.R., Kowalewski, D.E., *et al*. (2006) The age and origin of the Labyrinth, Western Dry Valleys, Antarctica: evidence for extensive Middle Miocene subglacial floods and freshwater discharge to the Southern Ocean. *Geology*, **34**, 513–16. [Photograph courtesy of: The Antarctic Photo Library, U.S. Antarctic Program].

6.3 MACROSCALE FEATURES OF GLACIAL EROSION

Macroscale features of glacial erosion are those features that are 1 km or greater in dimension. They are large enough to form significant landscape elements and may contain many of the smaller landforms already considered in this chapter. Five main landforms are recognised at this scale: (i) regions of areal scour; (ii) troughs; (iii) cirques; (iv) giant stoss and lee forms; and (v) tunnel valleys (Table 6.4).

6.3.1 Areal Scouring

The most commonly encountered landscape of glacial erosion is one of *areal scour*. It consists of an area of scoured bedrock composed of an assemblage of whalebacks, roches moutonnées, bedrock megagrooves and rock basins

Table 6.4   Macro-scale landforms of glacial erosion and their significance for the reconstruction of former ice masses

Landform	Morphology	Glaciological significance
Widespread areal scouring	Areas of low relief smoothed into streamlined eminences and basins, taking the form of roches moutonnées and whalebacks. Rock surfaces may contain striae, bedrock gouges and cracks. Detailed morphology controlled by preglacial weathering characteristics and patterns of bedrock jointing.	Warm-based ice carrying a basal debris load. Percentage of landscape accounted for by streamlined bedrock features and by stoss and lee features may indicate former basal conditions. Quarried landforms indicate low effective normal pressures (0.1–1 MPa) inferred from the presence of basal cavities with abundant basal meltwater and regular fluctuations in basal water pressure. Quarried landforms also indicate rapid sliding velocity and thin ice; direction and orientation of ice flow. Streamlined bedrock features and striae indicate high effective normal pressures (>1 MPa) inferred from intimate ice–bedrock contact and cavity suppression.
Glacial troughs Fjords	Deep valleys with smoothed, polished and steep walls and flat floors. Fjords are drowned glacial troughs. Valley cross-sectional morphology can be described by empirical power-law functions and by second-order polynomials.	Warm-based ice, abundant meltwater and high ice velocities. The cross-sectional area of a trough or fjord and its longitudinal profile may become calibrated to discharge over time, enabling estimates of palaeo-ice discharge to be made.
Cirques	Large bedrock hollows that open downslope and are bounded upslope by a cliff, steep slope or arcuate headwall.	Warm-based ice and abundant meltwater. Elevation and aspect of cirques commonly used in palaeoclimatic reconstructions to provide information on height of the former regional snowline and to assess the mass balance conditions under which empty cirques may become occupied.
Giant stoss and lee forms	Large upstanding bedrock hills or spurs with both abraded and quarried faces. Detailed morphology controlled by preglacial weathering characteristics and patterns of bedrock jointing.	Warm-based ice carrying a basal debris load. Low effective normal pressures (0.1–1 MPa) inferred from the presence of basal cavities. Quarried faces indicate abundant basal meltwater with regular fluctuations in basal water pressure. Rapid sliding velocity. Some evidence that roches moutonnées are deglaciation features formed under thin ice.

		May indicate direction and orientation of ice flow. Some features may indicate relatively low levels of glacial erosion, because they are commonly associated with preglacial valley spurs or bedrock hills.
Tunnel valleys and tunnel channels	Large, sinuous, steep-sided valleys or depressions that may contain enclosed basins in their floor. Tunnel channels are incised into bedrock, glacigenic sediment or other pre-existing materials. Tunnel valleys are usually infilled with sediment and occur both on the continental shelf and in lowland areas.	Related to subglacial meltwater discharge beneath large ice sheets, either via rapid drainage of stored subglacial meltwater or surface meltwater-derived drainage. Tunnel valleys and channels invariably indicate the presence of a melting ice sheet overlying a poorly consolidated substrate. Calculations of palaeovelocity and palaeodischarge possible from measurements of channel shape, channel width and size of material transported.

[Modified from: Glasser, N.F. and Bennett, M.R. (2004). *Progress in Physical Geography*, **28**, 43–75.]

(Figure 6.12A). Every part of the landscape is affected by glacial erosion, and depositional products are rare or absent (Figure 6.12B). The detailed morphology of the individual landforms within a region of areal scour is primarily controlled by the orientation, spacing and density of joints, foliations and other lines of weakness within the bedrock (Figure 6.12C and D). This type of landscape develops under extensive areas of warm-based ice. In Britain this type of terrain is sometimes referred to as *knock and lochan topography*. This term describes the upstanding rounded bedrock lumps (knocks) and the water-filled depressions (lochans) that separate them.

6.3.2 Glacial Troughs and Fjords

Glacial troughs and *fjords* (drowned troughs) are deep linear features carved into bedrock (Figure 6.13). They represent the effects of glacial erosion where ice flow is confined by topography and is therefore channelled along the trough or valley. Troughs may be cut beneath ice sheets as well as by valley glaciers and larger outlet glaciers. Glacial troughs are formed by a combination of both glacial abrasion and quarrying. Both these processes are required to produce steep-sided troughs, although the effects of glacial abrasion are more obvious due to the smoothed and polished trough walls. The processes of glacial quarrying are most evident on the valley floors, where quarried bedrock landforms such as roches moutonnées and rock basins are common.

Figure 6.12   Landscapes of areal scouring. (A) Areal scouring comprising whalebacks, roches moutonnées and rock basins in the Harlech Mountains of North Wales. (B) Ice-scoured bedrock dominated by ice-smoothed bedrock in Norway. Note person in centre of photograph for scale. (C) Glacially quarried bedrock blocks scattered around the lee side of a roche moutonnée in the Harlech Mountains of North Wales. The size of the blocks is directly related to the orientation and spacing of joints in the bedrock. (D) Ice-scoured bedrock on a hillside in western Ireland. Note how the glacial erosion has exploited the geological structure within the bedrock. [Photographs: N.F. Glasser]

The cross-sectional form of individual glacial troughs is often described as being 'U-shaped', but their true morphometry is more accurately described by empirical power-law or quadratic equations. If a slope profile is surveyed from the centre of a trough up one of its sides then this profile can be described by mathematical equations. In this way the cross-profiles of individual troughs can be compared to examine, for example, the role of lithology in determining trough morphology. The simplest equation used to describe the cross-sectional morphology of a trough is a power-law equation such as:

$$Y = aX^b$$

where Y is the vertical distance from the valley floor, X is the horizontal distance from the centre of the valley, a is a constant and b is a measure of the profile curvature.

Figure 6.13   Photograph of a glacial trough hanging above Golfo Elefantes, a deep fjord on the western coast of Chile. Note the steep valley sides and parabolic cross-section. [Photograph: N.F. Glasser]

Most glacial troughs have values of b of between 1.5 and 2.5. A parabola would have a value of 2. Alternatively the cross-sectional shape of a glacial trough can be described by a quadratic equation, such as:

$$Y = a + bX + cX^2$$

where Y is the vertical distance from the valley floor, X is the horizontal distance from the centre of the valley and a, b, c are coefficents determined stastically for each trough.

The choice of equation used in studies of troughs depends on the aim of the study. If the aim is to compare the variation of trough profiles from a standard shape such as the parabola then the power law equation is most applicable. However, if the aim is to compare the shape of individual troughs with one another then the quadratic equation is most appropriate.

The morphometric description of troughs is a powerful tool, because it allows the variation in trough form to be examined objectively. For example, we might expect trough morphology to vary with the lithology, or the strength of the rock mass, into which it is cut. By comparing mathematically the shape of troughs cut in one type of bedrock with those cut in another such hypotheses may be tested (Box 6.4).

It has been suggested that glacial troughs represent equilibrium landforms, such that once the initial morphology of a trough is established it changes little during erosion. This suggests that troughs represent the adaptation, by erosion, of

BOX 6.4: ROCK MASS STRENGTH AND TROUGH MORPHOMETRY

Working in the Southern Alps of New Zealand, Augustinus (1992) set out to test the hypothesis that the morphology of a glacial trough is related to the strength of the rock mass or bedrock into which it is cut. Determining the strength or resistance of bedrock to erosion is difficult because it is a function not only of the intact strength of the rock but also the density, spacing and orientation of the joints or other lines of weakness within the rock mass. To get around this problem, Augustinus used a method commonly utilised in studies of slope stability, known as the *Rock Mass Strength*, to estimate the geomorphological strength of the bedrock. The Rock Mass Strength classification involves scoring a rock mass in the field against a series of eight properties that collectively determine its strength. They are: (i) intact rock strength measured with a Schmidt hammer; (ii) the degree of weathering; (iii) the spacing between joints or partings; (iv) width of joints or partings; (v) the continuity of the joints or partings; (vi) the orientation of the joints or partings in relation to the slope; (vii) the presence or absence of infills along joints or partings; and (viii) the presence of the outflow of water from joints or partings. Augustinus (1992) determined trough morphometry and estimated the Rock Mass Strength of the rock mass into which each trough was cut at a series of different sample sites. The results indicate that there is a strong relationship between Rock Mass Strength and trough form. Glacial troughs appear to become narrower and the sides become steeper as the Rock Mass Strength increases. The number of troughs in a given area also appears to increase with Rock Mass Strength. This work illustrates neatly how the morphology of glacial troughs is partly controlled by the strength of the bedrock into which they are eroded.

Source: Augustinus, P.C. (1992) The influence of rock mass strength on glacial valley cross-profile morphometry: a case study from the Southern Alps, New Zealand. *Earth Surface Processes and Landforms*, **17**, 39–51.

preglacial valleys, where ice flow is difficult, into forms that are able to comfortably and efficiently accommodate ice flow. The amount of glacial erosion needed to create a glacial trough is therefore equal to the amount of adaptation needed to modify the preglacial valley in order to discharge the available ice efficiently. Once the shape of a glacial trough is established its size should be simply a function of the amount of ice that it has to discharge. This relationship has been confirmed by studies of outlet glaciers draining ice caps in Greenland and for valley glaciers in New Zealand, where there is a strong relationship between the drainage or accumulation area that supplies a glacier and the size of its trough. The relatively simple morphology of glacial troughs and their close relationship to ice discharge has also enabled their formation and evolution to be modelled numerically (Box 6.5). This type of analysis suggests that over time glacial valleys become adjusted to the glacial systems that form them. This might imply that most erosion will occur in

BOX 6.5: A NUMERICAL MODEL OF THE EVOLUTION OF A GLACIAL TROUGH

Understanding the formation of glacial troughs is difficult due to the lack of access beneath current glaciers and the length of time involved in their formation. Consequently, several researchers have tried to model the evolution of glacial troughs using computer models. Of particular note is the work of Harbor *et al.* (1988) who used a computer model to simulate the development of a glacial valley from an initial 'V-shaped' fluvial valley. They first modelled the pattern of glacier flow within a valley with 'V-shaped' cross-section and then used this to calculate the pattern of erosion within the valley as the ice flowed through it. Within the model the rate of erosion is assumed to be proportional to the square of the sliding velocity experienced by the glacier. After a period of time the new cross-sectional shape produced by the predicted erosion was calculated. This ice flow was then modelled through this new cross-section and the pattern of erosion predicted again. In this way the evolution of the valley's cross-sectional shape was modelled through a series of time steps and Harbor *et al.* (1988) were able to model the evolution of a 'U-shaped' glacial valley from an original 'V-shaped' valley cross-section. Their results are summarised in the diagram below and three stages of valley evolution were recognised:

Time 1: initial 'V-shaped' valley cross-section. During this period maximum erosion occurs at two points mid-way along the valley walls, producing a curved cross-section.

Time 2: intermediate stage. As the channel shape changes, the velocity and erosion patterns under the glacier are also changed. The two peaks of erosion reduce in magnitude and begin to shift towards the valley centre.

Time 3: final 'U-shaped' cross-section. At this stage erosion is concentrated at the base of the valley and the valley morphology remains relatively constant as the valley is incised into the landscape.

To reach this final steady-state requires the glacier to excavate a valley almost double the original valley depth. Harbor *et al.* (1988) calculated that this would take around 100 000 years given a rate of glacial erosion of around 1 mm per year and a final valley depth of 100 m.

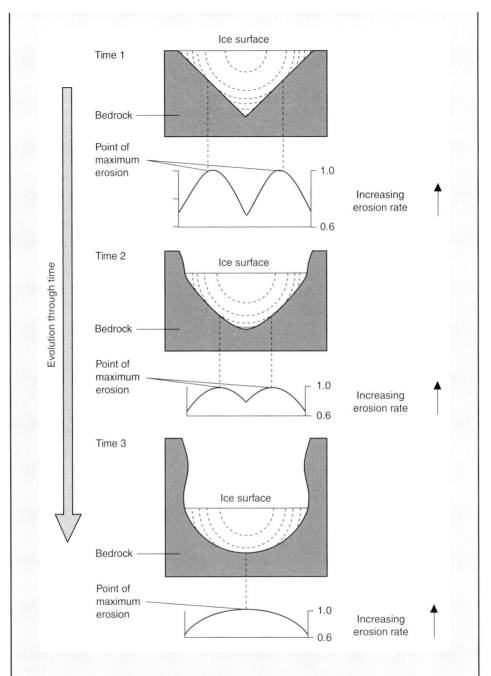

Source: Harbor, J.M., Hallet, B. and Raymond, C.F. (1988) A numerical model of land-form development by glacial erosion. *Nature*, **333**, 347–9.[Diagram modified from: Harbor *et al.* (1988) *Nature*, **333**, figure 1, p. 348]

an area when it is first glaciated, and once an efficient system of glacial troughs and other ice discharge routes has been established little modification may occur.

The longitudinal profile of glacial troughs usually contains a series of enclosed basins within it (Box 6.6). These are excavated as outlined in Section 6.2.3 through the operation of positive feedback between compressional flow and basin development. *Hanging valleys* are formed where two glacial troughs have been eroded into the landscape at different rates. If one trough is cut at a faster rate, perhaps because of a greater drainage area than its tributary, then the floor of the tributary will become perched at a higher altitude than that of the first trough. There may be a substantial height difference between the altitude of the two valley floors, and the highest is said to be left 'hanging' above the lowest.

BOX 6.6: NUMERICAL SIMULATIONS OF GLACIER VALLEY LONGITUDINAL PROFILES

Glacial erosion alters the cross-profiles of valleys by transforming fluvial valleys into much broader and deeper valleys (Box 6.7), but it is not clear how glaciations alter the longitudinal profile of valleys. To explore this, MacGregor *et al.* (2000) applied a climate-driven numerical model of glacial erosion to the development of the longitudinal profiles of glaciated valleys. Simulations of a single glacial valley showed rapid flattening of the long-itudinal profile. This is because sliding speed (which dictates the rate of glacial abrasion) and water-pressure fluctuations (which affects the rate of glacial quarrying) both peak at the equilibrium-line altitude (ELA), so that the most rapid erosion is focused around the ELA. As a development to the model, MacGregor *et al.* (2000) added a tributary glacier to their experiment. This addition resulted in the development of valley steps and overdeepenings immediately downvalley of the tributary junction, creating 'hanging' valleys above the main valley. These authors concluded that their model can therefore effectively reproduce the main components of glacial valleys.

Source: MacGregor, K.R., Anderson, R.S., Anderson, S.P. and Waddington, E.D. (2000) Numerical simulations of glacial-valley longitudinal profile evolution. *Geology*, **28**, 1031–4.

Valley patterns evolve from their preglacial arrangement with the intensity or duration of glaciation. Glacial erosion breaches preglacial watersheds (a process known as *divide elimination*) and new troughs may be formed to increase valley connectivity (Box 6.7). Modification of valley patterns therefore provides a crude

BOX 6.7: THE MODIFICATION OF VALLEY PATTERNS BY GLACIAL EROSION

Although we can easily identify individual landforms of glacial erosion it is often much more difficult to quantify the extent to which an entire landscape has been modified or affected by glacial erosion. Haynes (1977) and Riedel *et al.* (2007) quantified the effects of glacial erosion on the Scottish and the USA/Canadian Cordilleran landscapes respectively by examining the degree of valley connectivity in these areas. Fluvial landscapes tend to produce dendritic drainage patterns with low valley connectivity, whereas erosion by ice sheets will modify drainage patterns, breach watersheds and cut new troughs into the landscape. This will tend to increase the interconnectivity of the valley systems in a landscape. By using topological measures of connectivity, originally developed to study the connectivity of transport networks, these authors compiled maps to show the valley connectivity across the two areas. Connectivity is defined by two indices, α and β. The α index is defined by:

$$(E - V + G/2V - 5) \times 100$$

where V is the number of stream junctions, E the number of stream segments and G the number of separate sub-basins. The β index is defined by:

$$E/V$$

Alpha and beta values are determined from analysis of map data. A perfectly dendritic drainage network has an α value of 0 and a β value < 1. Diagram A shows the proportion of the terrain occupied by glacial valleys in Scotland, and Diagram B shows the connectivity of these valleys indicated by their α values. The maps show that valley connectivity, and therefore ice sheet erosion, is highest in north and west Scotland and lowest in the east and south. This pattern is to be expected since the high rates of accumulation and ablation on the more maritime west coast produce the steep mass balance gradients that favour intense glacial erosion. The lower mass balance gradients of the more continental eastern and southern parts of Scotland mean that these areas would be less likely to experience such intense erosion. This work provides a good example of a simple quantitative method by which we can determine the effects of glacial erosion on a landscape.

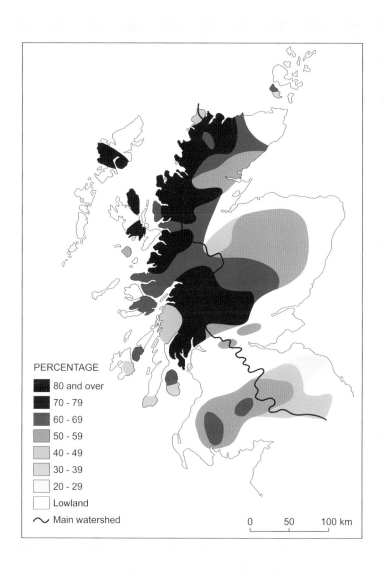

PERCENTAGE

- ■ 80 and over
- ■ 70 - 79
- ■ 60 - 69
- ■ 50 - 59
- ■ 40 - 49
- ■ 30 - 39
- □ 20 - 29
- □ Lowland
- ∿ Main watershed

0   50   100 km

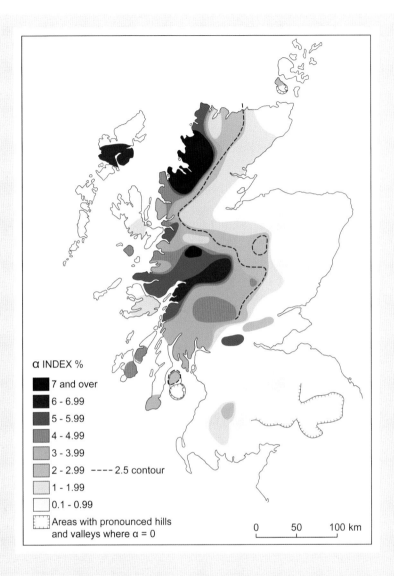

α INDEX %

- 7 and over
- 6 - 6.99
- 5 - 5.99
- 4 - 4.99
- 3 - 3.99
- 2 - 2.99   ---- 2.5 contour
- 1 - 1.99
- 0.1 - 0.99
- Areas with pronounced hills and valleys where α = 0

0   50   100 km

Sources: Haynes, V.M. (1977) The modification of valley patterns by ice-sheet activity. *Geografiska Annaler*, **59**A, 195–207. Riedel, J.L., Haugerud, R.A. and Clague, J.J. (2007) Geomorphology of a cordilleran ice sheet drainage network through breached divides in the North Cascades Mountains of Washington and British Columbia. *Geomorphology*, **91**, 1–18. [Modified from: Haynes (1977) *Geografiska Annaler*, **59**A, figures 3 and 4, p. 109]

guide to the intensity of glacial erosion. If the valley pattern is predominantly dendritic it retains its preglacial fluvial form and the intensity of erosion or duration of glacial erosion must have been low. On the other hand, if the dendritic pattern has been replaced by interconnected troughs then the intensity of glacial erosion has probably been much greater. The modification of valley patterns has been used to estimate the intensity of glacial erosion in Scotland and the North American Cordilleran (Box 6.7).

Fjords are drowned glacial troughs. These deep, often-linear features carved into bedrock represent the effects of glacial erosion in situations where ice flow is confined by topography and channelled along a trough or valley. Fjord landscapes are common in the Arctic and Antarctic, as well as along the maritime fringes of southern South America, northwest Europe and Norway, Canada, Alaska and New Zealand. Individual fjords are generally recognised to be palimpsest features, developed over successive glaciations. Fjords erode rapidly under glacial conditions and their considerable dimensions indicate that they represent significant volumes of rock removal by glacial erosion. Mean Quaternary glacial erosion rates of between 1 and 2 mm yr^{-1} have been calculated for glacial troughs in western Norway and Scotland. Valley patterns in fjord landscapes have been used to quantify the effects of glacial erosion and it has been suggested that the dimensions and longitudinal profiles of fjords are adjusted to the discharge of ice (Box 6.6).

6.3.3 Cirques

Cirques are large bowls that open downslope and are bounded upslope by a cliff or steep slope known as a *headwall*. The headwall is usually arcuate in plan and is much steeper than the cirque floor. The floor of the cirque may contain an enclosed rock basin and show evidence of glacial erosion, while the headwall is predominantly formed by glacial quarrying and periglacial freeze–thaw weathering. Cirques are usually created by individual glaciers, although they may drain collectively into larger valley glaciers.

Glacial cirques are found in mountainous terrain subject to local glaciation. They may also occur as part of landscapes of ice-sheet erosion, such as those in Britain (Figure 6.14). Here phases of local glaciation allowed cirques to develop, although their morphology has been modified during periods of more intense ice-sheet glaciation. Most cirques are the product of cumulative erosion during several phases of glaciation.

The precise definition of a cirque varies: for example in three separate studies of Scottish cirques the total number identified in the landscape varied between 347 and 876. Part of the reason for this is that cirques tend to grow in clusters, and individual features may become amalgamated over time to become composite features. For example, one large cirque may actually be a composite of several smaller feeder cirques. This variety of form makes the identification of individual cirques somewhat subjective. However, some of this subjectivity may be removed by using mathematical formulae to describe and classify cirque morphology.

Figure 6.14   Photograph of Cwm Cau, a glacial cirque cut into the mountain Cadair Idris in Snowdonia National Park, North Wales. [Photograph: N.F. Glasser]

In longitudinal profile, the headwall of a cirque is normally much steeper than its floor. The basic profile shape of a cirque therefore can be described by a logarithmic curve:

$$Y = k(1 - X)e^{-X}$$

where Y is the vertical distance from the valley floor, X is the horizontal distance from the centre of the valley, e is a constant and k is a shape constant.

This type of curve is known as a k-curve and the value of k is known as the k-number. For most cirques the k-number is between 0.5 and 2. The greater the value of the k-number the steeper the headwall of the cirque (Figure 6.15). A cirque fitted by a k-curve with a k-number of 2 will have a steep headwall and will be overdeepened to such an extent that its floor will mostly likely contain a lake. The type of long profile that a cirque possesses is a function of its bedrock lithology and of the structural weakness within it. For example, in areas where bedrock dips into a cirque then the floor of the cirque will also tend to dip inwards and the cirque may contain a small lake. Where bedrock dips out of the cirque it will have a floor that slopes outwards and will be best described by a low k-number (Figure 6.15). Jointing within the bedrock also determines the nature of the headwall. Closely spaced joints tend to produce blocky headwalls, whereas widely spaced joints produce smoother headwall profiles.

The evolution of cirque morphology through time can be examined using *ergodic reasoning*. Ergodic reasoning suggests that under certain circumstances sampling in space can be equivalent to sampling through time, and consequently, space–time transformation is permissible as a working tool. This works on the assumption that

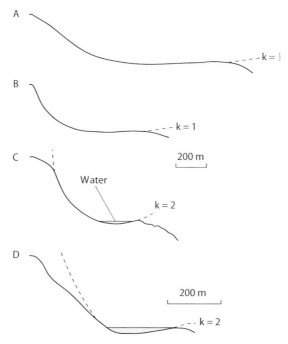

Figure 6.15 Empirical curves (*k*-curves) fitted to the longitudinal profiles of four Scottish cirques. The higher the value of *k*, the more enclosed the cirque basin. [Modified from: Haynes (1968) *Geografiska Annaler*, **50A**, figure 3, p. 223]

within a population of evolving landforms there will be a range of different examples at different stages of development. By comparing each example within a given area (i.e. in space) a model of landform evolution through time may be established. For example, within a mountainous area there may be a range of cirques each of which may have been initiated at a different time or developed at a different rate. Consequently the spatial variation in cirque morphometry within the area will provide insight into how cirques evolve through time.

This type of methodology has been used to establish the following models for the evolution of cirques in different areas. As cirques increase in size they appear to become more enclosed both in plan and in profile. Figure 6.16 shows a simple model of cirque evolution based on morphometric observation in the Scottish Highlands in which a well-defined cirque basin develops with time. In practice this type of model may hold only in areas of relatively uniform bedrock geology in which the structural weaknesses are present in all directions. In most rock masses this is not case and structural or lithological weaknesses tend to be orientated in one direction. In these situations the morphology of the cirque and its evolution may be strongly guided by orientations of these structural and lithological weaknesses. For example in the mountains of North Wales, cirques cut along the strike of the outcropping geology tend to be more elongated than those cut across strike. Work on these cirques also suggests that cirques which are orientated parallel to geological structures may evolve in a different fashion than those that are

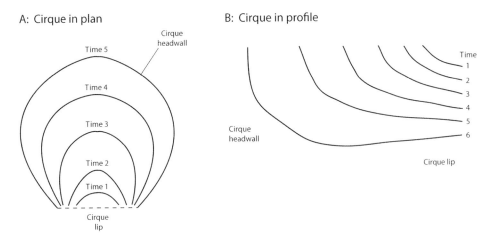

Figure 6.16   A model for the evolution of a cirque, based on observations of cirque morphology in the Scottish Highlands. (A) Cirque evolution in plan view. (B) Cirque evolution in profile view. [Modified from: Gordon (1977) *Geografiska Annaler*, **59A**, figure 7, p. 192]

orientated at right angles to the geological structure. Those cirques oriented along the geological structure appear to have experienced faster rates of headwall retreat than downcutting and consequently have evolved an elongated, flat-floored morphology. In contrast those that have developed transverse to the geological and lithological trend appear to have experienced similar rates of downcutting and headwall retreat. Consequently, they have evolved into more compact and enclosed basins. Once a basin has developed in the floor of a cirque it will continue to grow due the same positive feedback mechanism outlined for rock basins (see Section 6.2.3).

The elevation of cirques and their aspect can be used to provide general palaeoclimatic information. The altitude at which an abandoned cirque lies may be used as a measure of the former regional snowline within the area on the assumption that cirques are formed by discrete glaciers. In most cases this reflects a composite snowline averaged over numerous periods of glaciation during which the cirques were occupied. The closer the altitude of a cirque to sea level the lower the snowline. Cirques close to sea level provide evidence of relatively harsh climates, whereas those at high altitudes may form when the regional snowline is much higher. For example, cirque-floor altitudes within the Snowdon mountains of North Wales rise in elevation from the southwest to the northeast, a trend that has been interpreted as reflecting the direction of prevailing snow-winds during the later part of the Cenozoic Ice Age. The aspect of a cirque or direction in which they open is also indicative of palaeoclimate, because the location of the cirque glacier that cut the cirque is controlled by the direction of snow-bearing winds and the direction of incoming solar radiation. In the northern hemisphere most cirques face towards the northeast, although this aspect may be modified by strong snow-winds such that they form in the lee of mountain slopes crossed by the winds.

Cirques provide another good example of a positive feedback system because the deeper the cirque becomes the more efficient its shape is for trapping snow and shading the accumulation areas. The cirque glacier will consequently be bigger. A bigger glacier is likely to achieve more efficient erosion and therefore to continue to excavate the cirque, a cycle that will continue while the glacier survives.

6.3.4 Giant Stoss and Lee Features

Preglacial valley spurs and other hills may be eroded by ice into giant stoss and lee forms. These are given the collective term of *giant roches moutonnées* and range in size from hundreds of metres to several kilometres across. These features are carved in bedrock and may appear as either quarried valley spurs or as free-standing and isolated bedrock protrusions. Their morphology is similar to that of smaller roches moutonnées, with ice-smoothed proximal surfaces and quarried distal surfaces. This suggests that they are formed by a similar process but simply on a much larger scale. Smaller roches moutonnées and other erosional landforms are usually found superimposed on these asymmetrical spurs and hills.

Good examples of giant stoss and lee features are to be found in Glen Dee, Scotland, where large streamlined hills with lee-side cliffs up to 160 m high are to be found (Figure 6.17). These large-scale landforms formed under thin ice near the

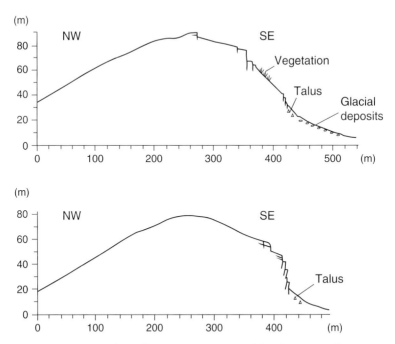

Figure 6.17 Cross-sections through two giant stoss and lee forms in Glen Dee, Scotland. [Modified from: Sugden et al. (1992) *Geografiska Annaler*, **74A**, figure 7, p. 259]

end of the last glaciation, during a time of high ice velocity and abundant melt-water. Other large stoss and lee features are to be found in south Greenland. These large composite landforms are ice-polished on their upstream side, whereas their downstream face is quarried. The upstream sides of these landforms contain smaller whalebacks superimposed upon the main landform.

6.3.5 Tunnel Valleys

Tunnel valleys are elongate depressions with overdeepened areas along their floors cut into bedrock or unconsolidated sediment. They are frequently sinuous in plan-form and may occur in anastomosing networks, although they also exist as inde-pendent, straight valleys. At their largest they may be 2 km wide, over 100 m deep and extend for between 6 and 30 km in length. In general, however, they tend to be smaller, perhaps only 600–800 m wide and less than 60 m deep. In long profile they contain enclosed hollows or isolated, often elongated, basins. They are commonly infilled to varying depths with a variety of different types of sediment, including sediment gravity flows and thick units of glaciofluvial sands. Tunnel valleys occur extensively on the continental shelf of northern Europe and North America, but buried examples have also been recorded on land. The morphological variation within tunnel valleys indicates three main theories of tunnel valley formation.

1. Tunnel valleys are cut into unconsolidated sediment where deformable subgla-cial sediment creeps into subglacial conduits from the sides and below, followed by removal of this material through the conduit by meltwater flow. Tunnel valleys are created, therefore, by lowering of the sediment surface on either side of the conduit.
2. Tunnel valleys form during deglaciation, at or close to the ice margin, by sub-glacial meltwater erosion and consequently the valleys are time-transgressive.
3. They form by subglacial meltwater erosion during high-discharge catastrophic channelised flood events. In this scenario, tunnel valleys within anastomosing networks would have formed synchronously.

6.4 LANDSCAPES OF GLACIAL EROSION

The landforms described above combine to produce landscapes of glacial erosion, which in turn provide important information about the dynamics and character of former glaciers and ice sheets. Basal thermal regime is particularly important because it is clear that warm-based ice, which slides over its bed, is much more erosive than cold-based ice, which is frozen to its bed (Box 6.8). Cold-based glaciers therefore achieve less glacial erosion than their warm-based counterparts (Box 6.9). Figure 6.18 shows the distribution of the landforms of glacial erosion in a cross-section through a hypothetical ice sheet. This conceptual model assumes that the ice is isothermal and that its velocity profile is simply a function of mass balance (see Section 3.5). The preservation of preglacial landforms is possible near the ice divide,

BOX 6.8: PRESERVATION OF LANDFORMS AND SEDIMENTS BENEATH COLD-BASED ICE SHEETS

In cold-based areas beneath ice sheets, where there is little or no basal meltwater, there is little or no glacier sliding. As a result, the basal layers of the ice sheet are frozen to the bed and the processes of glacial erosion are severely limited. This means that in these areas it is possible for older, non-glacial landforms to survive ice-sheet overriding. Evidence now exists for the preservation of a range of different landforms and sediments beneath former ice sheets. Initially these interpretations of landscape preservation were based solely on geomorphological evidence such as the survival of upland landscape elements, for example plateau surfaces (Clarhäll and Kleman, 1999). More recently, however, *cosmogenic radionuclide* evidence (Box 6.10) has been used to provide quantitative data on other landforms, which indicates that very old landscape elements can survive beneath ice sheets. Landforms dated in this way include upland tors (Stroeven *et al.*, 2002; Briner *et al.*, 2003) and moraines (Fabel *et al.*, 2006), as well as also lowland glacial features such as glaciomarine deltas (Davis *et al.*, 2006) and sediments in lake basins (Briner *et al.*, 2007). The photograph below shows a tor with a perched erratic in the Canadian Arctic from Briner *et al.* (2003).

Sources: Briner, J.P., Miller, G.H., Davis, P.T., *et al.* (2003) Last glacial maximum ice sheet dynamics in arctic Canada inferred from young erratics perched on ancient tors. *Quaternary science reviews*, **22**, 437–44. Briner, J.P., Axford, Y., Forman, S.L., *et al.* (2007) Multiple generations of interglacial lake sediment preserved beneath the Laurentide ice sheet. *Geology*, **35**, 887–90. Clarhäll, A. and Kleman, J. (1999) Distribution and glaciological implications of relict surfaces on the Ultevis Plateau, northwestern Sweden. *Annals of glaciology*, **28**, 202–8. Davis, P.T., Briner, J.P., Coulthard, R.D., *et al.* (2006) Preservation of Arctic landscapes overridden by cold-based ice sheets.

Quaternary research, **65**, 156–63. Fabel, D., Fink, D., Fredin, O., *et al*. (2006) Exposure ages from relict lateral moraines overridden by the Fennoscandian ice sheet. *Quaternary research*, **65**, 136–46. Phillips, W.M., Hall, A.M., Mottram, R., *et al*. (2006) Cosmogenic Be-10 and Al-26 exposure ages of tors and erratics, Cairngorm Mountains, Scotland: timescales for the development of a classic landscape of selective linear glacial erosion. *Geomorphology*, **73**, 222–45. Stroeven, A.P., Fabel, D., Hättestrand, C. and Harbor, J. (2002) A relict landscape in the centre of Fennoscandian glaciation: cosmogenic radionuclide evidence of tors preserved through multiple glacial cycles. *Geomorphology*, **44**, 145–54. [Photograph: J. Briner]

BOX 6.9: LANDFORMS OF EROSION AND DEPOSITION ASSOCIATED WITH COLD-BASED GLACIERS

Cold-based glaciers do not produce large quantities of subglacial meltwater, so it is commonly assumed that they neither slide nor abrade their beds. As a result, it is commonly assumed that they do not achieve much glacial erosion. Atkins *et al*. (2002) challenged this assumption, describing glacial geomorphological features associated with erosion and deposition by a cold glacier in the Allan Hills, Antarctica. They mapped and described a range of landforms, bedrock features and sediments including abrasion marks and striae, subglacial deposits, glaciotectonically deformed substrate, isolated blocks, ice-cored debris mounds and boulder trains. The photograph below, for example, shows a series of striations formed on a bedrock surface beneath a cold-based glacier in the Allan Hills, Antarctica. All of these features were inferred to have formed beneath a cold-based glacier as it advanced and receded. This study is important because it provides evidence that cold-based glaciers are capable of eroding, transporting and depositing material subglacially. The study also sets out some of the criteria for identifying the presence of former cold-based glaciers in the geological record.

Source: Atkins, C.B., Barrett, P.J. and Hicock, S.R. (2002) Cold glaciers erode and deposit: evidence from Allan Hills, Antarctica. *Geology*, **30**, 659–62. [Photograph: C. Atkins]

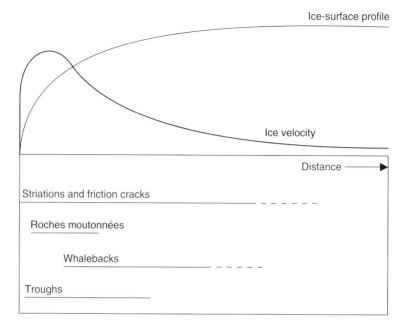

Figure 6.18   Schematic cross-section through a warm-based ice sheet showing the possible distribution of the landforms of glacial erosion.

where velocity is low and the ice may remain cold-based. Roches moutonnées develop where the ice is fast flowing and relatively thin, conditions that are ideal for cavity formation, whereas whalebacks occur only under thicker ice. Valley modification is greatest in the zone of fast-flowing ice close to the equilibrium line. One of the greatest advances in studies of glacial erosion in recent years is the use of *cosmogenic isotope ('exposure age') dating*, which can be used to test conceptual models such as this concerning the age of glacial landscapes, their evolution through time and their relationship to former ice dynamics and basal thermal regime (Box 6.10).

BOX 6.10: COSMOGENIC ISOTOPE DATING IN GLACIAL GEOMORPHOLOGY

Gosse and Phillips (2001) explain how terrestrial *in situ* cosmogenic nuclides (TCN) are used in geomorphology. Cosmogenic isotopes such as beryllium (^{10}Be) and aluminium (^{26}Al) are created when elements in the atmosphere or Earth are bombarded by cosmic rays that penetrate the atmosphere from outer space. The accumulation of these isotopes within a rock surface can be used to establish how long it has been exposed to cosmic radiation, rather than being shielded by a glacier or sediment. Exposures of surfaces from between a few thousand to about 10 million years old can be dated by the measurement of the ^{10}Be and ^{26}Al isotopes. The production rates of TCNs are almost unimaginably small – a few atoms per gram of rock per year. However, using accelerator mass spectrometry (AMS) and noble gas mass spectrometry it is possible to detect and count cosmogenic nuclides down to levels of a few thousand atoms per gram (parts per billion of parts per billion!). The build up of TCNs through time therefore provides a way to measure exposure ages and/or erosion rates for geomorphological surfaces directly from the mineral constituents of the surfaces. Consequently, the analyses of TCNs can be used to estimate 'exposure ages' for bedrock and boulder surfaces. This technique has revolutionised our ability to date periods of former glacial expansion because direct age estimates can now be obtained for landforms such as glacial moraines (by dating boulders on their surfaces), for ice-scoured bedrock (by dating striated bedrock surfaces) and for areas that remained ice-free above or beside former glaciers (for example by dating non-glacial surfaces above trimlines, on plateaux or on nunataks). A good example of this is the study of Briner *et al.* (2006) who used cosmogenic radionuclide measurements from bedrock samples and upland erratics from differentially weathered fjord landscapes on northeastern Baffin Island to shed light on Laurentide Ice Sheet dynamics and thickness. These authors found that tors on weathered upland surfaces have minimum ^{10}Be ages between ~ 50 and 170 ka ($n = 12$), whereas the majority of erratics perched in the uplands range from $\sim$10–13 ka ($n = 14$), indicating that the whole landscape was glaciated during the Last Glacial Maximum. They concluded that the uplands and lowlands were both covered by ice during the last glaciation but that the landscape was eroded differentially as a function of a spatially variable ice-sheet basal thermal regime.

Sources: Briner, J.P., Miller, G.H., Davis, P.T. and Finkel, R.C. (2006) Cosmogenic radionuclides from fiord landscapes support differential erosion by overriding ice sheets. *Geological Society of America Bulletin*, **118**, 406–20. Gosse, J.C. and Phillips, F.M. (2001) Terrestrial *in situ* cosmogenic nuclides: theory and application. *Quaternary Science Reviews*, **20**, 1475–560.

6.5 SUMMARY

Glacial erosion produces a range of landforms and landscapes. Landforms can be classified on the basis of scale. Typical microscale landforms include striations, micro crag and tails, friction cracks and p-forms. Mesoscale landforms comprise streamlined bedrock landforms, stoss and lee forms, grooves, rock basins and meltwater channels. Macroscale landforms include areal scouring, glacial troughs, cirques, giant stoss and lee forms and tunnel valleys. These landforms combine to produce landscapes of glacial erosion, which provide us with information about the dynamics and thermal characteristics of former glaciers and ice sheets.

SUGGESTED READING

Classifications of landforms of glacial erosion are presented by Laverdiere *et al.* (1979) and Dionne (1987). Glasser and Bennett (2004) outline how glacial erosional landforms can be used in palaeoglaciological reconstructions.

Abrasional bedrock forms such as glacial striations, micro crag and tails and friction cracks are described in papers by Gilbert (1906), Harris (1943), Dremanis (1953), Gray (1982), Shaw (1988) and Thorp (1981). Kleman (1990) and Iverson (1991) provide methodological discussions of the use of striations in reconstructing former ice-flow patterns and ice dynamics, respectively. Further examples can be found in Gemmell *et al.* (1986) and Lindstrom (1991). Sharp *et al.* (1989a) show how palaeoglaciological inferences can be made from microscale erosional landforms.

Classic papers on the origins of p-forms and glacial grooves are those of Dahl (1965) and Gjessing (1965). Gray (1981) describes some excellent examples in Scotland. Boulton (1974) argues that these features form by glacial abrasion, whereas Shaw (1994), Pair (1997) and Munro-Stasiuk *et al.* (2005) discuss their origin as meltwater landforms. Sharp *et al.* (1989b) discuss the significance and formation of microchannels cut by subglacial meltwater.

Streamlined bedrock features are discussed by Dionne (1987), Evans (1996), Bradwell (2005), Roberts and Long (2005) and Bradwell *et al.* (2008). A qualitative theory for the development of stoss and lee forms is presented by Carol (1947). The influence of bedrock structure on glaciated bedrock surfaces is described by Glasser *et al.* (1998) and on roche moutonnée morphology by Rastas and Seppala (1981) and Gordon (1981). Lindstrom (1988), Johansson *et al.* (2001) and Olvmo and Johansson (2002) consider the role of rock structure, lithology and preglacial deep weathering in determining the morphology of medium-scale glacial erosional landforms. More descriptions of the morphology of medium- and large-scale stoss and lee landforms are provided by Rudberg (1973), Glasser and Warren (1990) and Sugden *et al.* (1992).

The morphology and origin of meltwater channels are described in a range of papers, including Sissons (1960, 1961), Rodhe (1988), Sugden *et al.* (1991) and Greenwood *et al.* (2007). Studies of glacial troughs and fjords include the quantitative descriptions of Boulton (1974), Hirano and Aniya (1988), Harbor *et al*. (1988), Harbor (1992, 1995), Harbor and Wheeler (1992), Augustinus (1992a) and MacGregor *et al*. (2000). The role of ice discharge in creating glacial troughs is developed in papers by Haynes (1972) and Augustinus (1992b). A numerical model of fjord development is presented by Kessler *et al*. (2008). The role of ice sheets in changing large-scale drainage patterns is discussed by Haynes (1977) and Riedel *et al*. (2007). The relationship between hanging-valley heights and glacial erosion is described by Amundson and Iverson (2006).

Evans (2006) describes the global distribution of glacial cirques. The role of bedrock lithology and jointing in determining cirque morphology is described by Haynes (1968, 1995) and Gordon (1977). The morphology, distribution and possible origin of tunnel valleys are discussed by Ó Cofaigh (1996), Jorgensen and Sandersen (2006) and Hooke and Jennings (2006). Cosmogenic isotope dating has radically altered our understanding of the patterns and rates of glacial erosion, and information on applying this technique to landforms of glacial erosion can be found in Cockburn and Summerfield (2004), Fabel *et al*. (2004) and Li *et al*. (2005). The development of glaciated landscapes is discussed in papers by Glasser (1995), Glasser and Hall (1997), Sudgen *et al*. (2005), Briner *et al*. (2006, 2008) and Phillips *et al*. (2006). Finally, Boulton (1996), Bougamont and Tulaczyk (2003), Hildes *et al*. (2004) and Jamieson *et al*. (2008) provide examples of how numerical modelling can improve our understanding of glacial erosion and its links to sediment entrainment and transport at a landscape scale.

Amundson, J.M. and Iverson, N.R. (2006) Testing a glacial erosion rule using hang heights of hanging valleys, Jasper National Park, Alberta, Canada. *Journal of Geophysical Research-Earth Surface*, **111**, F01020.

Augustinus, P.C. (1992a) The influence of rock mass strength on glacial valley cross-profile morphometry: a case study from the Southern Alps, New Zealand. *Earth Surface Processes and Landforms*, **17**, 39–51.

Augustinus, P.C. (1992b) Outlet glacier trough size–drainage area relationships, Fiordland, New Zealand. *Geomorphology*, **4**, 347–61.

Boulton, G.S. (1974) Processes and patterns of glacial erosion, in (ed. D.R. Coates) *Glacial Geomorphology. Proceedings of the Fifth Annual Geomorphology Symposia, Binghampton*, Allen & Unwin, London, pp. 41–87.

Boulton, G.S. (1996) Theory of glacial erosion, transport and deposition as a consequence of subglacial sediment deformation. *Journal of Glaciology*, **42**, 43–62.

Bougamont, M. and Tulaczyk, S. (2003) Glacial erosion beneath ice streams and ice-stream tributaries: constraints on temporal and spatial distribution of erosion from numerical simulations of a West Antarctic ice stream. *Boreas*, **32**, 178–90.

Bradwell, T. (2005) Bedrock megagrooves in Assynt, NW Scotland. *Geomorphology*, **65**, 195–204.

Bradwell, T., Stoker, M. and Krabbendarn, M. (2008) Megagrooves and streamlined bedrock in NW Scotland: the role of ice streams in landscape evolution. *Geomorphology*, **97**, 135–56.

Briner, J.P., Miller, G.H., Davis, P.T. and Finkel, R.C. (2006) Cosmogenic radionuclides from fiord landscapes support differential erosion by overriding ice sheets. *Geological Society of America Bulletin*, **118**, 406–20.

Briner, J.P., Miller, G.H., Finkel, R.C. and Hess, D.P. (2008) Glacial erosion at the fjord onset zone and implications for the organization of ice flow on Baffin Island, Arctic Canada. *Geomorphology*, **97**, 126–34.

Carol, H. (1947) The formation of roches moutonnées. *Journal of Glaciology*, **1**, 57–9.

Cockburn, H.A.P. and Summerfield, M.A. (2004) Geomorphological applications of cosmogenic isotope analysis. *Progress in Physical Geography*, **28**, 1–42.

Dahl, R. (1965) Plastically sculptured detail forms on rock surfaces in northern Nordland, Norway. *Geografiska Annaler*, **47**, 83–140.

Dionne, J.C. (1987) Tadpole rock (rockdrumlin): a glacial streamline moulded form, in (eds J. Menzies and J. Rose) *Drumlin Symposium*, Balkema, Rotterdam, pp. 149–59.

Dremanis, A. (1953) Studies of friction cracks along the shore of Cirrus Lake and Kasakokwag Lake, Ontario. *American Journal of Science*, **251**, 769–83.

Evans, I.S. (1996) Abraded rock landforms (whalebacks) developed under ice streams in mountain areas. *Annals of Glaciology*, **22**, 9–16.

Evans, I.S. (2006) Local aspect asymmetry of mountain glaciation: a global survey of consistency of favoured directions for glacier numbers and altitudes. *Geomorphology*, **73**, 166–84.

Fabel, D., Harbor, J., Dahms, D., *et al.* (2004) Spatial patterns of glacial erosion at a valley scale derived from terrestrial cosmogenic Be-10 and Al-26 concentrations in rock. *Annals of the Association of American Geographers*, **94**, 241–55.

Gemmell, J., Smart, D. and Sugden, D. (1986) Striae and former ice-flow directions in Snowdonia, North Wales. *Geographical Journal*, **152**, 19–29.

Gilbert, G.K. (1906) Crescentic gouges on glaciated surfaces. *Geological Society of America Bulletin*, **17**, 303–13.

Gjessing, J. (1965) On plastic scouring and subglacial erosion. *Norsk Geografisk Tidsskrift*, **20**, 1–37.

Glasser, N.F. (1995) Modelling the effects of topography on ice sheet erosion, Scotland. *Geografiska Annaler*, **77**A, 67–82.

Glasser, N.F. and Bennett, M.R. (2004) Glacial erosional landforms: origins and significance for palaeoglaciology. *Progress in Physical Geography*, **28**, 43–75.

Glasser, N.F. and Hall, A.M. (1997) Calculating Quaternary glacial erosion rates in North East Scotland. *Geomorphology*, **20**, 29–48.

Glasser, N.F. and Warren, C.R. (1990) Medium scale landforms of glacial erosion in South Greenland: process and form. *Geografiska Annaler*, **72**A, 211–15.

Glasser, N.F., Crawford, K.R., Hambrey, M.J., *et al.* (1998) Lithological and structural controls on the surface wear characteristics of glaciated metamorphic bedrock surfaces: Ossian Sarsfjellet, Svalbard. *Journal of Geology*, **106**, 319–29.

Gordon, J.E. (1977) Morphometry of cirques in the Kintail–Affric–Cannich area of northwest Scotland. *Geografiska Annaler*, **59A**, 177–94.

Gordon, J.E. (1981) Ice-scoured topography and its relationships to bedrock structure and ice movement in parts of Northern Scotland and West Greenland. *Geografiska Annaler*, **63A**, 55–65.

Greenwood, S.L., Clark, C.D. and Hughes, A.L.C. (2007) Formalising an inversion methodology for reconstructing ice-sheet retreat patterns from meltwater channels: application to the British Ice Sheet. *Journal of Quaternary Science*, **22**, 637–45.

Gray, J.M. (1981) P-forms from the Isle of Mull. *Scottish Journal of Geology*, **17**, 39–47.

Gray, J.M. (1982) Un-weathered, glaciated bedrock on an exposed lake bed in Wales. *Journal of Glaciology*, **28**, 483–97.

Harbor, J.M. (1992) Numerical modelling of the development of U-shaped valleys by glacial erosion. *Geological Society of America Bulletin*, **104**, 1364–75.

Harbor, J.M. (1995) Development of glacial-valley cross sections under conditions of spatially variable resistance to erosion. *Geomorphology*, **14**, 99–107.

Harbor, J.M. and Wheeler, D.A. (1992) On the mathematical description of glaciated valley cross sections. *Earth Surface Processes and Landforms*, **17**, 477–85.

Harbor, J.M., Hallet, B. and Raymond, C.F. (1988) A numerical model of landform development by glacial erosion. *Nature*, **333**, 347–9.

Harris, S.E. (1943) Friction cracks and the direction of glacier movement. *Journal of Glaciology*, **51**, 244–58.

Haynes, V.M. (1968) The influence of glacial erosion and rock structure on corries in Scotland. *Geografiska Annaler* **50**A, 221–34.

Haynes, V.M. (1972) The relationship between the drainage areas and sizes of outlet troughs of the Sukkertoppen Ice Cap, West Greenland. *Geografiska Annaler*, **54**A, 67–75.

Haynes, V.M. (1977) The modification of valley patterns by ice-sheet activity. *Geografiska Annaler*, **59**A, 195–207.

Haynes, V.M. (1995) Alpine valley heads on the Antarctic Peninsula. *Boreas*, **24**, 81–94.

Hildes, D.H.D., Clarke, G.K.C., Flowers, G.E. and Marshall, S.J. (2004) Subglacial erosion and englacial sediment transport modelled for North American ice sheets. *Quaternary Science Reviews*, **23**, 409–30.

Hirano, M. and Aniya, M. (1988) A rational explanation of cross-profile morphology for glacial valleys and of glacial valley development. *Earth Surface Processes and Landforms*, **13**, 707–16.

Holmlund, P. (1991) Cirques at low altitudes need not necessarily have been cut by small glaciers. *Geografiska Annaler*, **73**A, 9–16.

Hooke, R.L. and Jennings, C.E. (2006) On the formation of the tunnel valleys of the southern Laurentide ice sheet. *Quaternary Science Reviews*, **25**, 1364–72.

Iverson, N.R. (1991) Morphology of glacial striae: implications for abrasion of glacier beds and fault surfaces. *Geological Society of America Bulletin*, **103**, 1308–16.

Jamieson, S.S.R., Hulton, N.R.J.H. and Hagdorn, M. (2008) Modelling landscape evolution under ice sheets. *Geomorphology*, **97**, 91–108.

Johansson, M., Olvmo, M. and Lidmar-Bergström, K. (2001) Inherited landforms and glacial impact of different palaeosurfaces in southwest Sweden. *Geografiska Annaler*, **83**A, 67–89.

Jorgensen, F. and Sandersen, P.B.E. (2006) Buried and open tunnel valleys in Denmark – erosion beneath multiple ice sheets. *Quaternary Science Reviews*, **25**, 1339–63.

Kessler, M.A., Anderson, R.S. and Briner, J.P. (2008) Fjord insertion into continental margins driven by topographic steering of ice. *Nature Geoscience*, **1**, 365–9.

Kleman, J. (1990) On the use of glacial striae for reconstruction of palaeo-ice sheet flow patterns. *Geografiska Annaler*, **72**A, 217–36.

Laverdiere, C., Guimont, P. and Pharand, M. (1979) Marks and forms on glacier beds: formation and classification. *Journal of Glaciology*, **23**, 414–16.

Li, Y.K., Harbor, J., Stroeven, A.P. *et al* (2005) Ice sheet erosion patterns in valley systems in northern Sweden investigated using cosmogenic nuclides. *Earth Surface Processes and Landforms*, **30**, 1039–49.

Lindstrom, E. (1988) Are roches moutonnées mainly preglacial forms? *Geografiska Annaler*, **70**A, 323–31.

Lindstrom, E. (1991) Glacial ice-flows on the islands of Bornholm And Christianso, Denmark. *Geografiska Annaler*, **73**A, 17–35.

MacGregor, K.R., Anderson, R.S., Anderson, S.P. and Waddington, E.D. (2000) Numerical simulations of glacial-valley longitudinal profile evolution. *Geology*, **28**, 1031–4.

Munro-Stasiuk, M.J., Fisher, T.G. and Nitzsche, C.R. (2005) The origin of the Western Lake Erie grooves, Ohio: implications for reconstructing the subglacial hydrology of the Great Lakes sector of the Laurentide Ice Sheet. *Quaternary Science Reviews*, **24**, 2392–409.

Ó Cofaigh, C. (1996) Tunnel valley genesis. *Progress in Physical Geography*, **20**, 1–19.

Olvmo, M. and Johansson, M. (2002) The significance of rock structure, lithology and pre-glacial deep weathering for the shape of intermediate-scale glacial erosional landforms. *Earth Surface Processes and Landforms*, **27**, 251–68.

Pair, D.L. (1997) Thin film, channelized drainage, or sheetfloods beneath a portion of the Laurentide ice sheet: an examination of glacial erosion forms, northern New York State, USA. *Sedimentary Geology*, **111**, 199–215.

Phillips, W.M., Hall, A.M., Mottram, R., *et al*. (2006) Cosmogenic [10]Be and [26]Al exposure ages of tors and erratics, Cairngorm Mountains, Scotland: timescales for the development of a classic landscape of selective linear glacial erosion. *Geomorphology*, **73**, 222–45.

Rastas, J. and Seppala, M. (1981) Rock jointing and abrasion forms on roches moutonnées, SW Finland. *Annals of Glaciology*, **2**, 159–63.

Riedel, J.L., Haugerud, R.A. and Clague, J.J. (2007) Geomorphology of a cordilleran ice sheet drainage network through breached divides in the North Cascades Mountains of Washington and British Columbia. *Geomorphology*, **91**, 1–18.

Roberts, D.H. and Long, A.J. (2005) Streamlined bedrock terrain and fast ice flow, Jakobshavns Isbrae, West Greenland: implications for ice stream and ice sheet dynamics. *Boreas*, **34**, 25–42.

Rodhe, L. (1988) Glaciofluvial channels formed prior to the last deglaciation: examples from Swedish Lapland. *Boreas*, **17**, 511–16.

Rudberg, S. (1973) Glacial erosion forms of medium size – a discussion based on four swedish case studies. *Zeit Schrift Für Geomorphologie*, **17**, 33–48.

Sharp, M., Dowdeswell, J.A. and Gemmell, J.C. (1989a) Reconstructing past glacier dynamics and erosion from glacial geomorphic evidence: Snowdon, North Wales. *Journal of Quaternary Science*, **4**, 115–30.

Sharp, M., Gemmell, J.C. and Tison, J. (1989b) Structure and stability of the former subglacial drainage system of the Glacier de Tsanfleuron, Switzerland. *Earth Surface Processes and Landforms*, **14**, 119–34.

Shaw, J. (1988) Subglacial erosional marks, Wilton Creek, Ontario. *Canadian Journal of Earth Science*, **25**, 1256–67.

Shaw, J. (1994) Hairpin erosional marks, horseshoe vortices and subglacial erosion. *Sedimentary Geology*, **91**, 269–83.

Sissons, J.B. (1960) Some aspects of glacial drainage channels in Britain, Part I. *Scottish Geographical Magazine*, **76**, 131–46.

Sissons, J.B. (1961) Some aspects of glacial drainage channels in Britain, Part II. *Scottish Geographical Magazine*, **77**, 15–36.

Sugden, D.E., Denton, D.H. and Marchant, D.R. (1991) Subglacial meltwater channel system and ice sheet over riding of the Asgard Range, Antarctica. *Geografiska Annaler*, **73**A, 109–21.

Sugden, D.E., Glasser, N.F. and Clapperton, C.M. (1992) Evolution of large roches moutonnées. *Geografiska Annaler*, **74**A, 253–64.

Sugden, D.E., Balco, G., Cowdery, S.G., *et al.* (2005) Selective glacial erosion and weathering zones in the coastal mountains of Marie Byrd Land, Antarctica. *Geomorphology*, **67**, 317–34.

Thorp, P.W. (1981) An analysis of the spatial variability of glacial striae and friction cracks in part of the Western Grampians of Scotland. *Quaternary Studies*, **1**, 71–94.

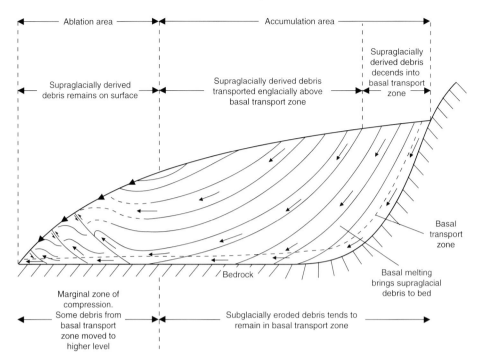

Figure 7.1   Debris transport pathways through a typical valley glacier. Two transport pathways can be identified: (i) a high-level pathway in which the debris does not come into contact with the bed; and (ii) a low-level pathway in which the debris is in contact with the bed. [Diagram reproduced with permission from: Boulton (1993) in *Holmes' Principles of Physical Geology* (ed. P.McL.D Duff), Chapman and Hall, figure 20.35, p. 425]

successive annual layers of accumulation (Box 7.1). The debris therefore follows an englacial path and will emerge from the glacier in the ablation zone to form a supraglacial cover as the ice surface is lowered by melting. Debris that accumulates

BOX 7.1: THE STRUCTURE OF GLACIERS

Glaciers contain a number of distinct structures. These can be divided into: (i) *primary structures* that result from accumulation, and: (ii) *secondary structures* that develop due to ice deformation during flow.

The principal primary structure is *ice stratification*, which results from the accumulation of snow each year. Summer surfaces are usually indicated by a refrozen melt-layer of bluish ice and by a concentration of debris that has fallen onto the glacier surface. *Regelation layers* are also primary structures, formed during the freezing of regelation ice to the base of the glacier (see Section 3.3.2). These primary structures become deformed during ice flow by internal deformation to produce secondary structures. The most important secondary structure is *ice foliation*, a layered fold structure of different sizes of ice crystals that develops by the deformation of primary ice structures during ice flow. It may develop either parallel to or transverse to the direction of glacier flow. *Longitudinal foliation* normally occurs at the ice margin and parallel to the direction of glacier flow. Most accumulation basins are wider than the channel that drains them; as a consequence of this, primary ice stratification is compressed and folded to form longitudinal folds or foliation. *Transverse* or *arcuate foliation* develops in regions with transverse crevasses or *ice falls*. Ice falls are densely crevassed regions of a glacier where ice descends a steep slope. Crevasses open as ice enters the ice fall, due to the extensional stress caused by flow acceleration in the ice fall. These crevasses then close at the base of the ice fall as extensional stresses change to compression. Former crevasses can be traced by the presence of different types of ice crystal that reflect light in different ways. These crevasse traces are deformed into an arcuate pattern by flow; extending further downstream in the centre of the glacier than at the sides. On some glaciers these crevasse traces become depicted by alternating bands of light and dark snow, known as *ogives*. It was suggested originally that each pair of dark and light ice bands represents a year's movement through the ice fall; the darker bands resulting from the concentration of dust and debris into crevasses during the summer months. However, their origin has now also been explained with reference to multiple shear zones in the ice, through which basal ice is uplifted to the glacier surface to produce the dark, foliated ogive bands. Finally, *thrusts*, sometimes referred to as *shear planes*, can develop in three situations: (i) in glaciers with a mixed or polythermal basal thermal regime; (ii) where glaciers flow against large subglacial bedrock obstacles or reverse bedrock slopes; and (iii) in surging glaciers.

Sources: Hambrey, M.J. and Lawson, W. (2000) Structural styles and deformation fields in glaciers: a review. *Geological Society of London Special Publications*, **176**, 59–83. Goodsell, B., Hambrey, M.J.and Glasser, N.F.(2002) Formation of band ogives and associated structures at Bas Glacier d'Arolla, Valais, Switzerland. *Journal of Glaciology*, **48**, 287–300.

on the glacier surface in the ablation zone will not be buried permanently by ice, unless it falls into a crevasse, and will therefore form a supraglacial cover.

Debris on the surface of a glacier may either be concentrated into down-glacier ridges known as *medial moraines* or will form an irregular layer over the glacier surface (Figure 7.2). There are two broad categories of medial moraine: *ice-stream interaction*; and *ablation-dominant*. Ice-stream interaction medial moraines are formed by the confluence of two lateral moraines at the junction of two glaciers (Figure 7.3). The medial moraine consists of a debris-covered ice ridge that extends down the trunk of the glacier and marks the line of suture between the two glaciers. It is important to note that the moraine is the surface expression of a vertical debris septum within the glacier that may extend to the glacier bed (Figure 7.3).

Figure 7.2   Vertical aerial photograph showing medial moraines at the confluence of two Svalbard valley glaciers. The glacier is about 1 km wide here and flow is from the bottom of the photograph towards the top. [Photograph: N.F. Glasser]

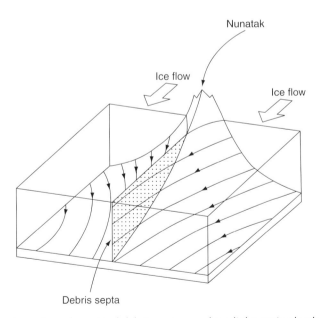

Figure 7.3   The formation of a vertical debris septum and medial moraine by the confluence of two glaciers.

Ablation-dominant medial moraines form where ridges of englacial debris are revealed down-glacier by surface melting. Such moraines appear to 'grow' out of the glacier surface in the ablation zone. This type of medial moraine may form in two ways.

1. If there is a point source for debris supply from a rockwall or nunatak onto the glacier, debris may take a linear configuration down-glacier of that point. In the accumulation zone this debris will slowly descend through the glacier in the direction of flow as it becomes buried by snow fall. In the ablation zone melting will reveal the debris as a medial moraine on the glacier surface.
2. Medial moraines can develop by the folding of debris-rich ice in longitudinal structures such as foliation and stratification. In the accumulation zone debris may collect as a diffuse layer over much of the glacier surface, where it will become buried by snow to form debris strata. Debris strata may become folded across the glacier during ice flow, particularly if the ice flows from a wide basin into a narrow trunk and is compressed laterally (Figures 7.4 and 7.5). The axes of these folds are usually parallel to the direction of ice flow. When the debris-rich ice reaches the ablation zone, surface melting lowers the glacier surface to reveal the anticlinal crests of the longitudinal debris-rich folds. The intensity of folding is determined by the amount of transverse compression within the glacier. If the folding is relatively open then a series of small medial moraines may emerge along the axis of each fold. However, if the folding is tight then the debris in the individual folds may merge to form a single medial ridge (Figure 7.4). It is important to recognise that the englacial and supraglacial debris structure of any glacier may be very complex and is a product of the cumulative deformation history of the ice.

The debris structure on the surface of a glacier is also affected by the rate of debris supply relative to the glacier's flow velocity. If the rate of debris supply is high and the glacier velocity is low, a thick layer of debris can accumulate. If the rate of debris supply is low and the glacier velocity is high, then the debris cover will be spread across the glacier surface more rapidly and it will be thinner. The presence of debris on a glacier surface has an important influence on its mass balance because supraglacial debris acts as insulation and slows down surface melting. Thick patches or ridges of supraglacial debris may be associated with large *ice-cored mounds*, *ridges* and *dirt cones* on the glacier surface. Debris-free ice melts rapidly on the clean ice surface, but is retarded beneath the debris cover itself. This process is particularly important in the development of ice-cored moraines, where blocks of glacier ice become buried beneath debris as the glacier retreats (see Section 9.1.3).

Debris transported at a high level within a glacier is often referred to as *passively transported* because it remains largely unaltered during glacial transport. It therefore retains its primary characteristics; it is typically angular, coarse and contains little in the way of fine material (Figure 7.6). In fact the sedimentological characteristics of debris transported at high levels within a glacier are often similar to those of talus or scree, which reflects the common origin of both as rockfall debris.

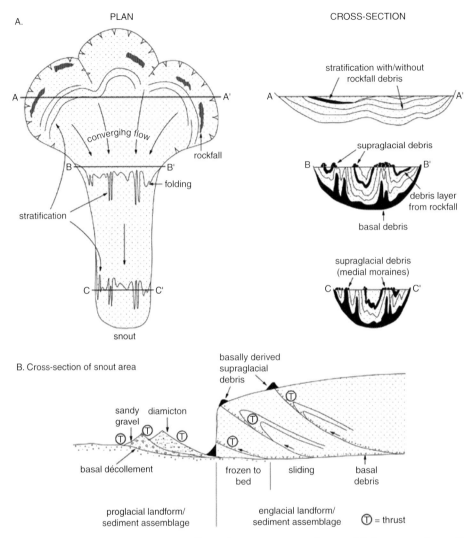

Figure 7.4 Schematic representation of how glacier structures influence debris transport in a valley glacier with multiple accumulation basins. (A) Plan view of a valley glacier showing the formation of medial moraines by folding within the glacier. (B) Cross-section showing development of thrusts near the glacier snout. [Modified from: Hambrey and Glasser (2005) in *Encyclopedia of Geology* (eds R.C. Selley, L.R.M. Cocks and I.R. Plimer), Elsevier, Amsterdam, figure 13, p. 673]

7.2 DEBRIS ENTRAINMENT

The evacuation and entrainment of rock fragments at the glacier bed were briefly outlined in Section 5.2.2 but before we consider low-level debris transport in depth we need to consider the main processes of *basal debris entrainment*.

Figure 7.5 Supraglacial debris stripes emerging from longitudinal foliation on the surface of Midre Lovenbreen, Svalbard. The glacier has receded in recent years and the debris stripes can be traced onto the glacier forefield. [Photograph: N.F. Glasser]

Figure 7.6 Photograph of supraglacial debris on the surface of a glacier in the Cordillera Blanca, Peru. Note that the debris is coarse, angular and contains little in the way of fine material. A supraglacial stream is beginning to rework the debris. [Photograph: N.F. Glasser]

1. **Regelation.** This process involves entrainment of debris at the bed by a combination of pressure melting and refreezing (regelation), to create a basal debris layer (Figure 7.7). It occurs under both temperate and polythermal glaciers. The matrix of this material is commonly fine-grained, comprising clay or silt. Clasts up to boulder-size, with subangular and subrounded shapes, faceted surfaces and striations, also occur.

Figure 7.7   The subglacial debris layer of a temperate glacier in Patagonia. Ice flow left to right. Note how the thickness of the basal debris layer changes as the glacier flows across the bed, particularly where it is compressed against the bedrock obstacle. Person in lower left indicates scale. [Photograph: N.F. Glasser]

2. **Entrainment associated with ice deformation.** Ice deformation can result in reorganisation of the debris that is already incorporated at the ice surface or bed. In polythermal glaciers in Svalbard, for example, three main modes of entrainment have been recorded in addition to regelation: (i) incorporation of angular rockfall material within the stratified sequence of snow, firn and superimposed ice; (ii) incorporation of debris of both supraglacial and basal character within longitudinal foliation; and (iii) thrusting, where debris-rich basal ice (including regelation ice) and subglacial sediments are uplifted into an englacial position, sometimes emerging at the ice surface.

3. **Entrainment associated with glaciohydraulic supercooling.** In recent years it has been suggested that supercooled meltwater flowing beneath glaciers is capable of entraining large quantities of sediment as it refreezes to the glacier sole on the upslope side of bed overdeepenings (see Section 2.4). As this debris-entrainment

process involves the refreezing of supercooled meltwater it is size-selective and favours the accretion of silt-size debris, which is typically present in meltwater. Typical ice facies include debris-rich frazil ice and anchor ice.

7.3 LOW-LEVEL DEBRIS TRANSPORT

Low-level debris transport is the transport of debris at or close to the base of the glacier. Debris may be derived directly from the bed by subglacial entrainment or indirectly from debris that falls onto the glacier surface and finds its way to the bed via crevasses and the downward movement associated with extending flow or basal melting (Figure 7.1). This debris remains at the base of the glacier until it is either deposited subglacially or released at the glacier snout or margin. Alternatively, it may be elevated onto the glacier surface along thrusts or by upward flow within the ice formed by compressive flow, a phenomenon common near the glacier snout.

Once entrained, basal debris becomes concentrated both laterally and vertically during glacier flow. The vertical extent of the basal debris may be increased by folding or thrusting of debris-rich basal ice. Debris can also become concentrated laterally around bedrock obstacles. As a uniform debris-rich basal ice

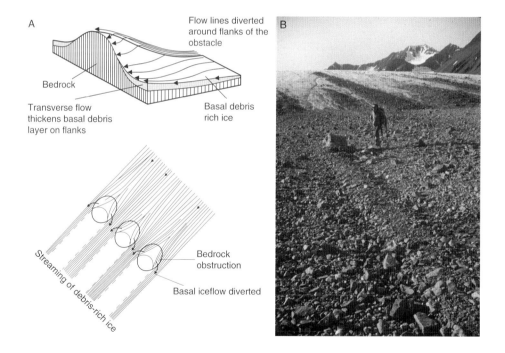

Figure 7.8 (A) The concentration of basal debris into distinct streams around subglacial bedrock obstacles. (B) A typical flute in the lee of a boulder on a glacier forefield in Svalbard. Former ice flow was towards the camera. (A) Modified from: Boulton (1974) in *Glacial Geomorphology* (ed. D.R. Coates), George Allen and Unwin, figure 12, p. 62. (B) Photograph: N.F. Glasser

layer approaches a bedrock obstacle its flow accelerates due to enhanced creep around the sides of the obstacle, such that less debris-rich ice is carried over the top of the obstacle (Figure 7.8). This mechanism explains how it is possible to develop a complex and laterally dispersed basal sediment layer.

The basal transport zone can be divided into two subzones: (i) a *zone of traction* in which debris is moved along the bed of the glacier; and (ii) a *zone of suspension* in which debris is transported immediately above the glacier base. The transfer of debris between these two zones is controlled primarily by pressure melting and regelation. Pressure melting on the up-ice side of an obstacle causes particles to move downwards, while the freezing-on of ice and debris on the down-ice side causes debris to move upwards. Folding, particularly against bedrock steps, may also cause particles to move upward from the zone of traction.

Debris transported at a low level within a glacier is often referred to as *actively transported* because it is altered during glacial transport. Particles in transport within the zone of traction experience considerable modification through the processes of crushing and abrasion (*communition*). They are typically spherical and rounded, and usually have a bimodal or multimodal grain-size distribution. The grain-size distribution is typically composed of three separate populations: (Figure 7.9) (i) large rock particles, *lithic fragments*; (ii) *mineral grains* produced by crushing of the rock fragments; and (iii) *submineral-sized particles* produced by the abrasion of mineral grains. In an ideal environment in which debris is neither removed nor added, the relative importance of these two populations should change as the transport distance increases: the fine mineral population should grow at the expense of the coarse population (Figure 7.9; Box 7.2). In practice this is rarely observed because of the constant addition of new material.

Roundness should also increase with transport distance as the corners of a particle are blunted and smoothed off by particle abrasion. However, observations suggest that roundness does not increase indefinitely, but reaches a *terminal roundness*. This reflects the fact that rock particles are also crushed subglacially as they are transported, which increases their angularity, at the same time as they are abraded. The degree of roundness a particle can achieve will be controlled by the length of time between crushing events, which will depend on its strength and the force applied. The stronger or more resistant a particle is to crushing, the more rounded it may become. Particles transported at the base of glacier are also characterised by faceted and striated surfaces, which often give clasts a bullet-shaped appearance (Figure 7.10).

Particles in transport within basal ice also develop a strong *particle fabric*, as elongated particles become aligned with the direction of ice flow. A preferred orientation develops parallel to the direction of flow although subsequently particles may also become orientated in a direction transverse to flow. This is because the orientation that provides the least resistance to flow is the one in which the long axis is parallel to the direction of flow. This property is of particular importance because glacial sediments may inherit this particle fabric.

Material can also be transported under a glacier by subglacial meltwater streams and through sediment deformation (e.g., as part of a deforming bed or during glaciotectonism). Figure 7.11 represents an attempt to summarise these low-level debris entrainment and transport mechanisms.

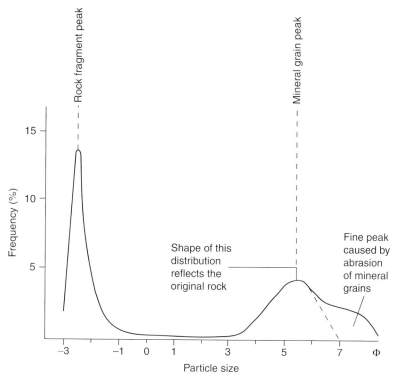

Figure 7.9   The grain-size distribution of subglacially transported debris is composed of three distinct populations: (i) lithic or rock fragments; (ii) mineral grains produced by crushing of the rock fragments; (iii) fines produced by the abrasion of individual mineral grains.

BOX 7.2: PARTICLE-SIZE DISTRIBUTIONS AND TRANSPORT DISTANCES

Dreimanis and Vagners (1971) argued on the basis of crushing experiments that the relative magnitude of the two particle-size peaks of subglacial sediment should change with transport distance. As transport distance increases, the finer mode (composed predominantly of mineral grains) should grow at the expense of the coarser mode (composed of predominantly lithic fragments). Dreimanis and Vagners (1971) demonstrated this by examining the particle-size distributions in three till samples in the Hamilton Niagara region of south-east Canada. The particle size of only one clast lithology (dolomite) was analysed in order to keep the results relatively simple. The results, illustrated below, show that with increased transport distance the finer mode increases in importance, although the amount of transport necessary will depend on the lithology. The patterns are of course more complex when multiple lithologies are considered and where new material is introduced to the glacier by debris entrainment. More recently, Hooke and Iverson (1995) have argued that particle-size distributions can be used to distinguish between those sediments that have been subjected to subglacial sediment deformation and those that have not.

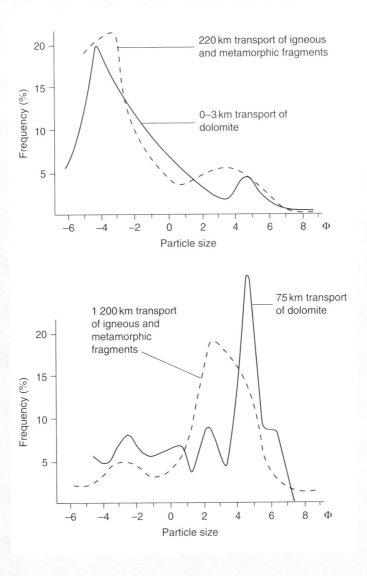

Source: Dreimanis, A.and Vagners, U.J. (1971) Bimodal distribution of rock and mineral fragments in basal tills, in *Till: a Symposium* (ed. R.P. Goldthwait), Ohio State University Press, Ohio, pp. 237–50. Hooke, R.L. and Iverson, N.R. (1995) Grain-size distribution in deforming subglacial tills: role of grain fracture. *Geology*, **23**, 57–60. [Modified from: Dreimanis and Vagners (1971) in *Till: a Symposium* (ed. R.P. Goldthwait), Ohio State University Press, Ohio, figure 4, p. 243].

Figure 7.10   A subglacially transported clast showing faceting and striations [Photograph: N.F. Glasser]

7.4 DEBRIS TRANSFER BETWEEN LOW AND HIGH LEVELS

7.4.1 Glacier Structures

An understanding of *glacier structures* or *structural glaciology* is vital to understanding how debris is distributed and transferred within a glacier (Box 7.1). These structures are principally the product of internal deformation, and are intimately associated with debris transport (Figure 7.12). Glacier ice is similar to any other type of geological material in that it comprises strata that progressively deform to produce a wide range of structures. *Primary structures* include sedimentary stratification derived from snow and superimposed ice, unconformities and regelation layering resulting from pressure melting and refreezing at the base of a glacier. *Secondary structures* are the result of deformation, and include both brittle features (crevasses, crevasse traces, faults and thrusts) and ductile features (foliation, folds, boudinage). Typically, a glacier will reveal a sequential development of structures over time and consequently ice at the glacier snout may record several phases of deformation. These phases of deformation reflect the passage of a 'parcel' of ice through the glacier from top to bottom. Ice high in the accumulation area will therefore be experiencing its first phase of deformation while ice at the snout may be experiencing its third or fourth phase.

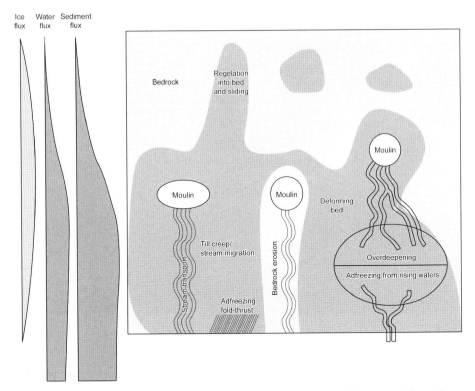

Figure 7.11   Plan view showing a schematic configuration of the possible sediment transport mechanisms beneath a glacier. The glacier is flowing from the top towards the bottom of the page. [Modified from: Alley *et al.* (1997) *Quaternary Science Reviews,* **16,** figure 2, p. 1031]

7.5 DEBRIS TRANSFER

Debris transfer occurs to some extent at most glacier snouts in the ablation area because here compressive flow causes the upward flow of debris-rich ice. Within some glaciers this process is facilitated by the development of *thrusts, thrust faults* or *shear planes,* which may transfer basal debris into an englacial, or supraglacial location. Thrusts develop in three situations: (i) in glaciers with a mixed or polythermal basal thermal regime; (ii) where glaciers flow against large subglacial bedrock obstacles or reverse bedrock slopes; and (iii) in surging glaciers.

Thrusts are most common in glaciers with a mixed thermal regime. The zone of thermal transition from warm- to cold-based ice is associated with strong compression. The warm-based ice slides over its bed, but the cold-based ice does not. This causes a down-glacier deceleration in velocity, which can generate compression beyond that which ice-creep can accommodate. As a consequence of this, a fault (thrust) may develop. These thrusts form at the base of the glacier and allow basal

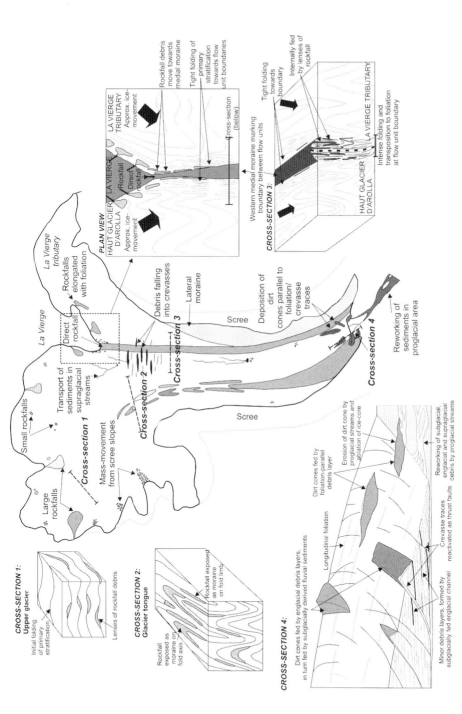

Figure 7.12 Schematic summary of contemporary debris transport processes at Haut Glacier d'Arolla, Switzerland. The plan view shows the main inputs and transport processes occurring on the glacier surface. Cross-sections 1–3 summarise medial moraine formation. Cross-section 4 shows the relationship between glacier structures and debris deposition near the glacier snout. [Reproduced with permission from: Goodsell *et al.* (2005) *Journal of Glaciology*, **51**, figure 9, p. 144]

Figure 7.13   Thrusts exposed in a lateral cliff at the glacier Midre Lovenbreen, Svalbard. Ice flow is right to left. Numerous low-angle thrusts can be seen rising from the bed towards the ice surface. [Photograph: N.F. Glasser]

ice to move up and over the colder ice in front (Figure 7.13). Thrusts may penetrate throughout the thickness of a glacier or terminate englacially as *blind thrusts*. Thrust planes often exploit structural weaknesses within the ice such as the traces of former crevasses. They may occur singularly and extend considerable distances across an ice margin (Figure 7.13), or in anastomosing networks in which each thrust has a short arcuate outcrop on the surface of the glacier. The angle between the glacier bed and the thrust plane is generally low (30–40°), although high-angle thrusts (40–80°) have been documented in surge-type glaciers. Thrusting is a particularly important mechanism within surging glaciers.

Debris entrainment into thrusts occurs in a variety of different ways, varying from incorporation of thin layers of debris-rich ice to large rafts of frozen sediment that are frozen to the glacier bed and elevated along thrust planes (Figure 7.14). Debris-rich thrusts may penetrate the full depth of a glacier so that the debris crops out on the glacier surface, or they may terminate englacially (Figure 7.14). Meltwater can also rise along thrusts, elevating substantial quantities of glaciofluvial sediment to the ice surface (Figure 7.15).

The transfer of basal debris into an englacial or supraglacial position is particularly important in determining the depositional processes that operate within a glacier (see Section 8.1). The presence or absence of thrusts, for example, is an important control on its debris structure. Thick englacial and supraglacial debris concentrations can be created by thrusting and these can have important implications for the release of debris from the glacier. Thrusts are comparatively rare in

Figure 7.14 The terminal zone of a Svalbard valley glacier, showing debris arranged on the ice surface in mounds and pinnacles as a result of ice-structural control. The debris has been elevated along thrusts within the ice. [Photograph: N.F. Glasser]

Figure 7.15 A pinnacle of debris on the surface of Marthabreen, Svalbard. This photograph illustrates neatly how different populations of sediment can occur in close proximity, with well-rounded cobbles and boulders melting out of a thrust and mixing with the supraglacial material on the ice surface. [Photograph: N.F. Glasser]

warm-based glaciers and thick basal debris layers are unusual because basal melting prevents their development. As a result, debris transport in temperate glaciers is dominated by high-level (passive) transport. In contrast, polythermal glaciers often have well-developed thrust planes. The result is that these glaciers have thicker englacial debris concentrations and the supraglacial debris at the glacier snout contains a mixture of debris transported at both high and low levels. The subglacial debris layer may also be thicker due to regelation in zones of thermal transition between warm- and cold-based ice. Glaciers that are completely cold-based throughout are dominated by high-level debris transport, although subglacial debris layers can develop.

7.6 SUMMARY

There are two principal transport pathways for debris within a glacier: (i) a high-level transport route (supraglacial and englacial) where transport is passive; and (ii) a low-level or basal pathway (subglacial) where transport is active. The debris transport pathway determines the properties of the resulting glacial debris because low-level transport involves modification of the debris during transport, whereas high-level transport does not. Basal debris typically contains more fines and more rounded and spherical clasts than debris that does not come into contact with the glacier bed. Basal debris often has a multimodal or bimodal grain-size distribution composed of a coarse mode of lithic fragments and a finer mode of mineral-sized grains. The long axis of debris in basal transport tends to become preferentially orientated in the direction of ice flow. The characteristics of debris transported at high and low levels within a glacier are sufficiently different for them to be used by glacial geologists to infer the transport history of glacial debris. This is particularly important in the analysis of ancient glacial deposits (Box 7.3). In general terms, polythermal glaciers tend to carry a high basal debris load, and their surfaces rarely

---

BOX 7.3: DETERMINING THE TRANSPORT HISTORY OF ANCIENT GLACIAL DEBRIS

In north eastern Svalbard there is a well preserved sequence of tillites – lithified till – of Late Precambrian age. These tillites are of uncertain origin. Dowdeswell *et al.* (1985) used the typical sediment characteristics of modern glacial sediments to interpret these ancient sediments. In particular they used particle fabric (orientation), but also made reference to particle shape. On the basis of particle fabric they were able to suggest that most of the tillites were either flow tills or glaciomarine diamicton (Box 8.2). From the particle shape analysis they were able to determine that most of the clasts within the tillites had undergone some form of basal transport, being typically well-rounded, striated and spherical. The relative absence of clasts with supraglacial characteristics led them to suggest that the supply of supraglacial sediment to the ice sheet which deposited the tillites in the Late Precambrian may have been very small. This would suggest that the glacier which deposited them had the dimensions of an ice cap or ice sheet, as one would expect a much higher proportion of supra-glacial clasts if the tillites had been deposited in an ice-field or by a valley glacier over looked by large supraglacial debris source areas. The work illus-trates how simple sedimentological observations of particle shape can be used to make important inferences about ancient ice sheets.

Source: Dowdeswell, J.A., Hambrey, M.J. and Wu, R. (1985) A comparison of clast fabric and shape in Late Precambrian and modern glacigenic sediments. *Journal of Sedimentary Petrology*, **55**, 691–704.

have a substantial cover of debris. In contrast, temperate glaciers, especially those in alpine terrain, normally carry little debris at the bed, but supraglacial debris is commonly scattered over their surfaces. As we will see in subsequent chapters, the resulting sedimentary products can be used to infer the thermal and topographic regimes of former ice sheets and glaciers.

SUGGESTED READING

Alley *et al.* (1997) review the physical processes that govern glacial debris entrain-ment and transport and Knight (1997) describes the basal layers of glaciers. Iverson (1993, 2000) outlines theoretical and experimental observations of debris entrain-ment at the glacier bed by regelation. Cuffey *et al.* (2000) explain debris entrainment under cold-based glaciers while Larson *et al.* (2006) make the case that glaciohy-draulic supercooling may have been an important debris entrainment mechanism beneath former ice sheets. Boulton (1978) quantifies the differences between those glacial sediments that have been in contact with the bed and those that have not. Ballantyne (1982) builds on this work at contemporary glaciers in Norway, and Benn (1989) provides an example of debris transport by former glaciers. Changes in basal particle characteristics with transport are discussed by Humlum (1985) and Kjaer *et al.* (2003). Eyles and Rodgerson (1978), Small (1987), Anderson (2000) and Hambrey and Glasser (2003) provide useful information about the formation of medial moraines. Structural glaciology is reviewed by Hambrey and Lawson (2000) and the role of structural glaciology in debris entrainment, transport and deposition in polythermal glaciers is outlined in a series of papers by Bennett *et al.* (1996a,b), Glasser *et al.* (1998, 2003), Hambrey *et al.* (1997, 1999) and Hubbard *et al.* (2004). Goodsell *et al.* (2005) provide an example of debris transport in a temperate alpine valley glacier. Methods for determining the transport history of both ancient and modern glacial sediments are given in Dowdeswell *et al.* (1985), Benn and Ballantyne (1994) and Bennett *et al.* (1997). The development of particle fabric within ice is discussed in the classic paper by Glen *et al.* (1957).

8: Glacial Sedimentation on Land

Glaciers terminating on land can deposit sediment in two ways: (i) directly, that is from the ice itself; or (ii) indirectly, for example through the action of meltwater. Direct deposition by a glacier is restricted to its immediate vicinity, whereas meltwater may carry sediment far beyond the glacier margin. Glacial sediments are traditionally divided into those that are non-sorted and formed by direct glacier sedimentation, and those that are sorted or stratified and usually deposited by meltwater. In practice this is a rather artificial distinction because meltwater deposits are not always stratified. In this chapter we describe the products of direct glacier sedimentation before considering deposition from meltwater.

8.1 DIRECT GLACIAL SEDIMENTATION

Debris deposited directly by a glacier is known as *till*. A till is defined as sediment deposited by glacier ice, but one which has not been disaggregated, although it may have suffered glacially induced flow either in the subglacial or supraglacial environment. It normally consists of large pebbles, cobbles or boulders, referred to generally as *clasts*, set within a fine-grained matrix of silt and clay. Its characteristics are, however, highly variable, as eloquently expressed in the following statement 'till is a sediment and is perhaps more variable than any sediment known by a single name' (Flint, 1957, p. 105). The definition of till is important because it excludes situations where the sediment has settled through water, such as is the case where a glacier terminates in a lake or in the sea. Some authors have used the term waterlain till for till-like sediments deposited in such environments, but it is perhaps more appropriate to use the non-genetic term diamicton. *Diamicton* is a non-sorted or poorly sorted unconsolidated sediment that contains a wide range of particle sizes for which no genesis is presumed; all tills are diamictons, but not all diamictons are tills (Figure 8.1). The term *diamict* is preferred by some authors, although technically it is a collective term for a diamicton and *diamictite*, the latter being a lithified diamicton.

Glacial Geology: Ice Sheets and Landforms   Second Edition   Matthew R. Bennett and Neil F. Glasser
© 2009 John Wiley & Sons, Ltd

Figure 8.1   Close up of a subglacial till or diamict. [Photograph: M. R. Bennett]

There are four primary processes by which debris in transport within a glacier may be deposited: (i) *lodgement* – this occurs when the frictional resistance between a clast in transport at the base of a glacier and the glacier bed exceeds the drag imposed by the overlying ice such that the clast ceases to move; (ii) *meltout* – the direct release of debris by melting; (iii) *sublimation* – vaporisation of ice causing the direct release of debris; and (iv) *subglacial deformation* – this involves the assimilation of sediment into a deforming layer beneath a glacier. The process of sublimation is currently only documented from Antarctica and occurs there due to the extreme cold and aridity of this environment.

Traditionally these process distinctions are used to recognise a range of different till types on the basis of their macromorphology, namely (Table 8.1): (i) *lodgement till*; (ii) *subglacial meltout till*; (iii) *deformation till*; (iv) *supraglacial meltout* and *flow till*; and (v) *sublimation till*. However, this distinction has increasingly been challenged over the past 10 years with the development of micromorphological analysis of till. Micromorphology involves the collection of undisturbed sediment blocks from the field, which are then impregnated with resin and thin-sectioned for study under a microscope. Careful analysis of the internal architecture of these sediments suggests that the traditional classifications are no longer valid and that most tills form by a combination of processes, especially in the subglacial environment. Distinguishing a lodgement till from a deformation till for example is not possible at a microscopic scale, because both show evidence of deformation. In fact it has recently been argued that tills should be referred to as *tectomicts* – the product of a complex suite of tectonic rather than depositional processes. Although such terminology has yet to be widely adopted it does draw attention to the problem of distinguishing lodgement till from meltout and deformation till in the subglacial environment. To address this issue here we recognise two basic till domains: (i) *subglacial tills*; and (ii) *supraglacial tills*.

Table 8.1 Summary table of the main sedimentary characteristics of the main types of till.

	Lodgement till	Subglacial meltout till	Deformation till
Particle shape	Clasts show characteristics typical of basal transport: rounded edges, spherical form, and striated and faceted faces. Large clasts may have a bullet shaped appearance.	Clasts show characteristics typical of basal transport, being rounded, spherical, striated and faceted. These characteristics are less pronounced than those of lodgement till.	Dominated by the sedimentary characteristics of the sediment that is being deformed, although basal debris may also be present.
Particle size	The particle-size distribution is typical of basal debris transport, being either bimodal or multimodal.	The particle-size distribution is typical of basal debris transport, being either bimodal or multimodal. Sediment sorting associated with dewatering and sediment flow may be present.	Diverse range of particle sizes reflecting that found in the original sediment. Rafts of the original sediment may be present causing marked spatial variability.
Particle fabric	Lodgement tills have strong particle fabrics in which elongated particles are aligned closely with the direction of local ice flow.	Fabric may be strong in the direction of ice flow, although it may show a greater range of orientations than that typical of lodgement till.	Strong particle fabric in the direction of shear, which may not always be parallel to the ice-flow direction. High-angle clasts and chaotic patterns of clast orientation are also common.
Particle packing	Typically dense and well consolidated sediments.	The sediment may be well packed and consolidated, although this is usually less marked than in a lodgement till.	Densely packed and consolidated.
Particle lithology	Clast lithology is dominated by local rock types.	Clast lithology may shows an inverse superposition.	Diverse range of lithologies reflecting that present within the original sediments.
Structure	Massive structureless sediments, with well-developed shear planes and foliations. Sheared or brecciated clasts, smudges, may be present. Boulder clusters or pavements may occur within the sediment along with evidence for ploughing of clasts.	Usually massive but if it has been subject to flow it may contain folds and flow structures. Crude stratification is sometimes present. The sediment does not show evidence of shearing and overriding during formation.	Fold, thrust and fault structures may be present if the level of shear homogenisation is low. Rafts of undeformed sediment may be included. Smudges (brecciated clasts) may also be present.

Table 8.1 Continued.

	Supraglacial meltout (moraine) till	Flow till	Sublimation till
Particle shape	Usually dominated by sediment typical of high-level transport, but subglacially transported particles may also be present. The majority of clasts are not normally striated or faceted.	Broad range of characteristics, but dominated by particles that are angular and have a non-spherical form. The majority of clasts are not striated or faceted.	Clasts typical of basal transport being rounded, spherical, striated, and faceted.
Particle size	The size distribution is typically coarse and unimodal. Some size sorting may occur locally where meltwater reworking has occurred.	The size distribution is normally coarse and unimodal, although locally individual flow packages may be well sorted.	The particle size distribution is typical of basal debris transport, being either bimodal or multimodal.
Particle fabric	Clast fabric is unrelated to ice flow, is generally poorly developed and spatially highly variable.	Variable particle fabric. Individual flow packages may have a strong fabric, reflecting the former palaeoslope down which flow occurred.	Strong in the direction of ice flow, although it may show a greater range in orientation than a typical lodgement till.
Particle packing	Poorly consolidated, with a low bulk density.	Poorly consolidated with a low bulk density.	Typically has a low bulk density and is loose and friable.
Particle lithology	Clast lithology is usually very variable, and may include far-travelled erractics.	Variable, but may include far-travelled erractics.	Clast lithology may show an inverse superposition.
Structure	Crude bedding may occur but generally it is massive and structureless.	Individual flow packages may sometimes be visible. Crude sorting, basal layers of tractional clasts may be visible in some flow packages. Sorted sand and silt layers may be common, associated with reworking by meltwater. Individual flow packages	The deposit is usually stratified and may preserve englacial fold structures.

8.1.1 Subglacial Till

Till may accumulate in a subglacial setting via a range of processes, including: (i) direct lodgement of debris in traction over the glacier bed; (ii) basal melting and debris release; (iii) deposition in basal cavities; and (iv) deformation and assimilation of overridden sediment.

1. **Direct lodgement.** In order to understand this process it is best to first consider a single clast in transport at the base of a glacier. A clast in transport at the base of a glacier need not move forward at the same speed as the basal ice: the ice may flow around the particle as it transports it. The particle will lodge or stop moving when its forward velocity is reduced to zero. This will occur whenever the friction between the particle and the bed exceeds the drag on the particle provided by the ice flowing over it. At this point the ice will simply flow around the particle without moving it forward. A particle may, therefore, lodge beneath flowing ice. Figure 8.2 shows several ways in which the velocity of an individual particle or mass of particles may be reduced to zero as they move over a rigid bed or plough through a soft one. In a simple abrasion model such as that proposed by Geoffrey Boulton (see Section 5.1.4; Figure 5.2) lodgement is part of a continuum, with erosion controlled by effective normal pressure. In this simplified scenario lodgement is controlled by variables such as: (i) an increase in ice thickness, which will increase the effective normal pressure; (ii) a fall in basal water pressure; and (iii) a fall in ice velocity.

2. **Subglacial melting.** Sediment is supplied to the sole of the glacier and released by basal melting. It may then be transported subglacially before it finally lodges. The rate of basal melting is determined by: (i) the geothermal heat flux; (ii) the amount of frictional heat generated by sliding and ice deformation, which increases with ice velocity towards the equilibrium line; (iii) ice thickness, because increasing the thickness of a glacier may increase basal ice temperature; (iv) the rate of advection of cold or warm snow; and (v) ice-surface temperatures. This is discussed in more detail in Section 3.4.

3. **Cavity deposition.** Subglacial cavities are known to occur where glaciers flow over irregular bedrock surfaces. Large cavities may form in the lee of bedrock obstacles, especially where the ice is thin and fast-flowing. Sediment can accumulate within these cavities in a variety of ways (Figure 8.3).

4. **Subglacial deformation.** Sediments can form a mobile deforming layer beneath glaciers flowing over soft substrates (see Section 3.3.3; Box 3.4). This layer may consist of soft preglacial sediment – either non-glacial or earlier glacial sediment – which has been overrun and deformed or it may consist of sediment deposited in a syn-sedimentary context by subglacial meltwater or by lodgement and meltout processes. The deformation of this sediment by subglacial shear may give rise to a homogeneous glacial till, even if it starts out as something very different. Think for a moment of mixing red jam into white yogurt; initially the two are distinct but as one mixes (i.e. one applies stress) the jam is drawn out into layers that are progressively folded into the yogurt until finally a homogeneous

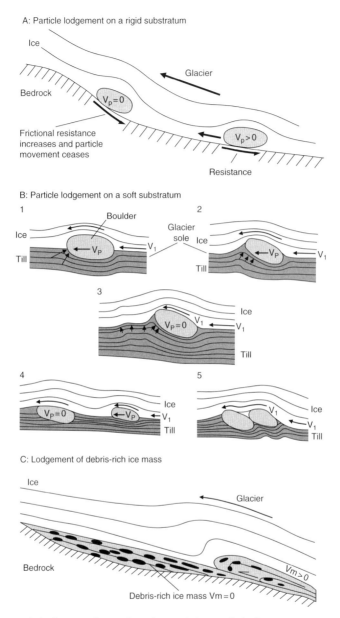

Figure 8.2 Particle lodgement beneath a glacier. (A) Particle lodgement on a rigid substratum. The clast in transport stops moving when the frictional resistance between it and the bed exceeds the drag imposed on the clast by the flowing ice. (B) Particle lodgement on a soft substratum. Clasts plough through soft sediment and will be stopped when the sediment ploughed up in front of them provides sufficient resistance to retard forward movement. Subsequently other clasts may jam against the first, and in this way boulder pavements or concentrations may form. (C) Lodgement of a debris-rich ice mass. A debris-rich ice body may lodge beneath a glacier when the frictional resistance between it and the bed exceeds that of the ice above, which shears over the debris-rich ice mass. [Modified from: Boulton (1982) in: *Research in Glacial, Glacio-Fluvial and Glacio-Lacustrine Systems* (eds R. Davidson-Arnott, W. Nickling and B.D. Fahey), Geo Books, figure 2, p.6]

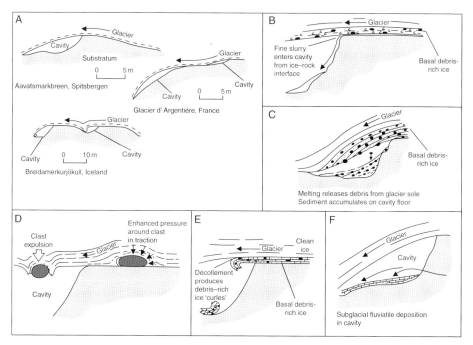

Figure 8.3 Observed mechanisms by which debris accumulates on the floor of subglacial cavities. [Modified from: Boulton (1982) in: *Research in Glacial, Glacio-Fluvial and Glacio-Lacustrine Systems* (eds R. Davidson-Arnot, W. Nickling and B.D. Fahey), Geo Books, figure 1, p. 4]

pink mixture results. Figure 8.4 illustrates this point in a geological context. *Glaciotectonic deformation* takes place whenever the stress imposed by a glacier exceeds the strength of the material beneath or in front of it. The material may be subject to both brittle (faults, thrusts) and ductile (folds) deformation depending on the pore-water pressure within the sediment. Ductile deformation is favoured by high pore-water pressures, which reduce the internal friction or strength of the material allowing it to deform. Deformation of this sediment proceeds in stages depending upon the amount of shear stress applied. At low levels of shear the sediment is simply folded and faulted. As the level of shear increases these structures slowly become attenuated and the nose of folds may be detached from their core, or *derooted*, to create *boudins*. Boudins are sausage-shaped blocks of less ductile material surrounded by a more ductile medium. In time they may become attenuated and drawn out at high levels of shear to form *tectonic laminations* (Figure 8.4). Sediments that experience very high-levels of shear become completely mixed and homogenised. The product of intense deformation is therefore a homogeneous diamict in which all the original sedimentary structures of the deposit have been destroyed. Some authors use the term *glacitectonite* for a sediment that retains some of the structural characteristics of the material from which it is derived after deformation; that is a sediment that has not been completely remoulded and homogenised.

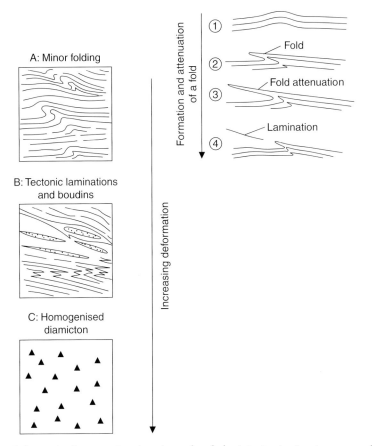

Figure 8.4   Schematic diagram showing the style of glaciotectonic structures associated with different levels of subglacial deformation. The insert shows how a fold may be attenuated to form a tectonic lamination. [Modified from: Hart and Boulton (1991) *Quaternary Science Reviews*, **10**, figure 14, p. 347]

Most subglacial sediment owes its origin to a combination of the above processes, although deformation beneath moving ice is probably the dominant process. Recognition of distinct subtypes of till is therefore not possible. This message has been repeatedly reinforced in recent years by micromorphological work, which reveals that most till contains a complex record of superimposed deformation events of varying styles (brittle versus ductile) and that the process history is therefore best interpreted in terms of a tectonic framework. The range of material incorporated into subglacial tills is one of the reasons for the diverse nature of this sediment. In some cases it may be dominated by glacially derived sediments produced by erosion and the comminution of clasts during transport. In other cases it may simply reflect the tectonic mixing of preglacial sediments under subglacial shear.

Given the importance of subglacial deformation to the formation of subglacial till it is worth exploring a little further the spatial and temporal controls on this process. Some of the fundamental principles of subglacial deformation are listed below.

1. Subglacial sediment will deform whenever the applied stress exceeds the strength of the material. The strength of the material is determined by the pore-water pressures within it; high pore-water pressures allow the grains to move more easily past one another and reduces the internal friction within the sediment. In sediment with low pore-water pressures, deformation will tend to occur in a brittle fashion along distinct failure surfaces such as fractures or faults, whereas at high pore-water pressure the same sediment may deform in a ductile fashion via a series of folds. Deformation will also occur along the weakest horizon within sediments with properties that are not homogeneous. It is also worth noting that as granular sediments deform they may expand or dilate, and that this dilated sediment deforms at a lower shear stress than that initially necessary to overcome the internal friction within the sediment at rest. Dilatancy will also control how easily pore-water can drain from sediment as stress varies. Finally, sediment properties may be modified during deformation, with changes to grain-size distributions due to clast-to-clast crushing, as well during the assimilation of new material into the deforming layer.

2. Sediment properties vary spatially and temporally beneath a glacier due to spatial facies variations across the glacier bed and temporal variations in subglacial hydrology. As a consequence not all areas of a glacier bed will be in motion at the same time due to deformation. Instead, it is better to envisage a mosaic of deforming and non-deforming (*sticky spots*) patches below a glacier. The distribution of these sticky spots may vary in time and will of course have an impact on the overall velocity component of a glacier due to subglacial deformation.

3. Sediment within a deforming layer will move forward under the applied stress, a process sometimes referred to as *till advection*. If the flow of deformable sediment into a given area equals the out flow of sediment from that area then deposition will not occur unless the geometry or dynamics of the glacier changes. However, if more sediment flows into an area than out, for example from an area of rapid transport to one of little transport, then sediment accumulation will occur. Alternatively if one considers a glacier with extending or accelerating ice flow downstream, then downcutting will occur because output will exceed the input of sediment. In these areas deformation will be evidenced only by a sharp basal slip or *décollement* surface. In contrast, in areas of compressional flow due to decreasing basal shear stress down-ice, subglacial deformation will also decrease down-ice and sediment accumulation may occur because output will be less than the input. The point is illustrated in Figure 8.5 where a

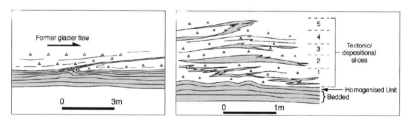

Figure 8.5 Progressive accumulation of a deformation till via individual tectonic slices in a compressive or deaccelerating flow regime. [Modified from: Boulton *et al.* (1991) *Quaternary International*, **86**, figure 2, p. 8]

series of attenuated tectonic laminations accumulate as a series of slices, progressively one above the other. For an ice sheet flowing over a deforming bed of uniform character this relationship between extending and compressing flow will result in a pattern of erosion and deposition by subglacial deformation like that shown in Figure 8.6.

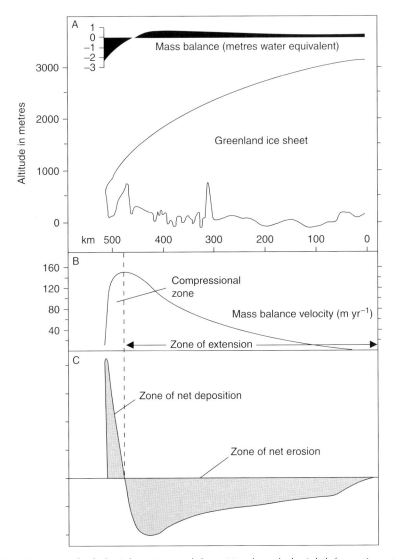

Figure 8.6 Patterns of subglacial erosion and deposition by subglacial deformation within the Greenland Ice Sheet. [Diagram reproduced from: Boulton (1987) in: *Drumlin Symposium* (eds J. Menzies and J. Rose), Balkema, figure 5, p. 37. Copyright © 1987, Taylor & Francis]

Deposition of a subglacial till involves the immobilisation of the deforming layer and is therefore essentially controlled by changes in applied stress as well as by the material properties of the deforming layer. It is important to note, however, that not all workers agree with the pervasive nature of subglacial deformation and some suggest that the properties of deformation may result simply from the ploughing of boulder and ice keels at the ice–sediment interface (Box 8.1). It is also possible that subglacial meltout may occur without deformation, for example below stagnant ice that has ceased to move. In these special situations debris may meltout sufficiently undisturbed to retain characteristics of the basal ice from which it is derived. These processes were once thought to be quite widespread, however, situations where ice melts out without being deformed either by movement or by the overburden are probably very limited and may not have good preservation potential.

BOX 8.1: WERE DEFORMING GLACIER BEDS WIDESPREAD?

During the early 1990s there was a paradigm shift within glacial geology which involved the increasing acceptance of the idea of subglacial deformation as a key glacial process responsible not only for many subglacial tills but also for associated landforms such as drumlins. Not all researchers have accepted the idea that glacier bed deformation was extensive beneath former mid-latitude ice sheets. At the heart of this question lies a debate over the interpretation of the glacial evidence and about how sediment actually deforms. Subaerial sediments, such as those found on hill slopes, tend to fail in a plastic fashion as explained by classic soil mechanics. These sediments have a finite strength known as the yield strength, above which they fail. The rate of subsequent deformation – strain rate – is independent of the stress applied. Sediment that behaves in this way tends to fail along a discrete horizon or surface. There is one school of thought which argues that this type of model best explains what goes on beneath a glacier. This view has been supported by observation beneath the Whillans Ice Stream in Antarctica and by laboratory based studies. If this model is correct then deformation will occur in a very thin zone and the development of a thick deforming layer is unlikely. Promoters of this model suggest that thick deforming layers seen in the geological record reflect the ploughing of ice bed keels and boulders through sediment along with the cumulative effects of multiple failures. Piotrowski *et al.* (2001) argued eloquently for this model and identified a range of sedimentological observations that favour this interpretation. They suggested that bed deformation was consequently restricted beneath former mid-latitude ice sheets.

The alternative is that subglacial sediments fail via some form of non-linear viscous rheology in which strain rate is dependent on the applied stress. Consequently deformation will occur through a greater thickness of sediment rather than along a discrete horizon, thereby giving a thicker deforming layer.

This type of model is consistent with the original observations of subglacial deformation reported in Box 3.4 and with the development of a thick layer of deforming sediment.

Hindmarsh (1997) reconciles these two schools of thought rather elegantly. He acknowledged, as most researchers now do, that at a small-scale till fails in a plastic fashion. During such events the rate of deformation is independent of the stress regime. He suggests, however, that the net integration of multiple small-scale plastic failures is best approximated by a viscous flow law. He draws the analogy with ice, which on an atomic scale behaves in a manner similar to plastic deformation, but the net effect of individual crystal dislocations is non-linear viscous type behaviour. He argues that a non-viscous approach explains the geological products of subglacial deformation such as patterns of erosion, deposition and the formation of glacial bedforms such as drumlins. Boulton *et al.* (2001) responded to the points raised by Piotrwoski *et al.* (2001), arguing for widespread bed deformation and demonstrating how similar glacial sequences can be interpreted in very different ways.

Source: Boulton, G.S., Dobbie, K.E. and Zatsepin, S. (2001) Sediment deformation beneath glaciers and its coupling to the subglacial hydraulic system. *Quaternary International*, **86**, 3–28. Hindmarsh, R.C.A. (1997) Deforming beds: viscous and plastic scales of deformation. *Quaternary Science Reviews*, **16**, 1039–56. Piotrowski, J.A., Mickelson, D.M., Tulaczyk, S., *et al.* (2001). Sediment deformation beneath glaciers and its coupling to the subglacial hydraulic system. *Quaternary International*, **86**, 3–28. Piotrowski, J.A., Mickelson, D.M., Tulaczyk, S., *et al.* (2001) Were deforming subglacial beds beneath past ice sheets really widespread? *Quaternary International*, **86**, 139–50.

One potential exception to this is in cold arid environments, such as Antarctica, where ablation may occur by sublimation. Sublimation is the direct vaporisation of ice without it passing through a liquid phase and consequently ice removal may be achieved without disturbing the debris to the same degree as more conventional meltout. Sublimation can occur in both subglacial and supraglacial locations, but the focus has been on subglacial till formation via this mechanism (sublimation till; Table 8.1). For post-depositional deformation to be avoided, sublimation must occur below stagnant ice. Sublimation till may preserve some of the debris structure inherited from the ice, such as stratification and englacial fold structures.

8.1.2 Supraglacial Till

Melting on the surface of a glacier, driven by solar radiation, releases debris that can produce supraglacial meltout till (Figure 8.7). This debris may be confined to debris

Figure 8.7   Supraglacial debris on a Himalayan glacier. [Photograph: N.F. Glasser]

transported at a high-level within the glacier or alternatively may incorporate debris from basal ice if it is elevated to the surface at the glacier snout (see Section 7.4).

As the debris accumulates on the glacier surface it first accelerates melting, because dark surfaces absorb more heat than reflective ones, but then insulates the surface from further melting as the debris thickness increases. Variation in the thickness of the debris causes variation in insulation and therefore variation in the surface-ice topography. The debris becomes concentrated into ridges and mounds of buried ice. This debris is unstable and prone to slumping and surface redistribution. Due to the almost constant movement of debris on the glacier surface it rarely retains any of the characteristics of the debris-rich ice from which it is derived. Scree-like characteristics, crude bedding and downslope clast fabrics may develop due to the flow or fall of material down ice-cored slopes. Several different facies of supraglacial moraine till may be identified, depending on the thickness of the original supraglacial debris cover and upon the level of fluvial reworking (Figure 8.8). The debris on a glacier surface is commonly concentrated into ice-cored ridges, which may trace the outcrop of debris-rich structures on the glacier surface. The characteristics of a supraglacial meltout till are summarised in Table 8.1. The clast content is usually dominated by sediment typical of high-level transport (see Section 7.1), although basal debris may also be present where it has been elevated by flow compression at a glacier snout. Consequently, clasts show a broad range of characteristics, but are dominated by angular particles, with a non-spherical form and few surface striations. The size distribution is also typically coarse and unimodal with a clast fabric that is

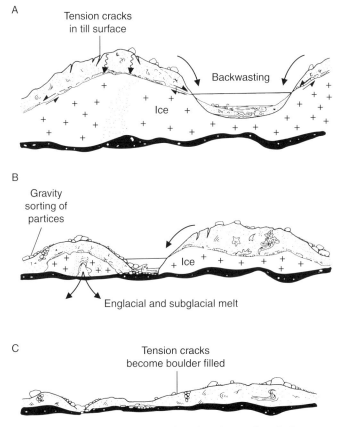

Figure 8.8   Progressive development of supraglacial meltout till with down-wasting of buried ice. Note the complex range of processes involved and the constant reworking of the sediment. [Modified from: Eyles (1979) *Canadian Journal of Earth Science,* **16**, figure 5, p. 1348]

unrelated to ice flow, and generally poorly developed, although locally strong fabrics may occur where sediment has fallen or slumped, scree-like, down ice-cored slopes. Supraglacial meltout tills are also typically poorly consolidated with a low bulk density.

Supraglacial debris is often saturated and located on a constantly changing ice surface due to surface ablation. As a result, supraglacial meltout till is highly unstable and consequently subject to downslope flow. Sediment that has flowed in this way is know as 'flow till' and its properties will depend on the water content, the debris character, the surface gradient and whether the surface over which the debris is moving is composed of ice or debris (Figure 8.9). In general, the greater the water content the more fluid the debris flow (Figure 8.10). There are many types of mass movement that can re-mobilise supraglacial debris although it is possible to recognise three main types.

Figure 8.9 Large debris flow lobe (flow till) on the surface of Kongsvegen, Svalbard. [Photograph: M.R. Bennett]

1. **Mobile flows.** These are thin, highly fluid, rapid flows, which are erosive and show crude size-sorting with coarse particles tending to settle to the base of the flow. The particles are usually strongly orientated in the direction of flow.

2. **Semi-plastic flows.** These are thick and slow moving tongues of debris, which are erosive. They may show size-sorting, with coarse particles settling to the base of the flow, and their upper surfaces may be sorted by the flow of meltwater. Fold structures and a weak particle orientation may develop. These types of flows usually result from the failure and flow of the downslope edge of sediment at the base of a laterally retreating ice-cored slope.

3. **Creep.** Slow, downslope movement of debris, which is not visible to the naked eye. This may occur either as a general non-channelised lobe of debris or as a more-or-less continuous sheet of creeping mass. Particles are rarely orientated in the direction of flow.

In practice, although it is possible to recognise different types of flow on modern glaciers, ancient flow-till deposits simply consist of numerous *flow packages*, of varying type stacked one on top of the other. Consequently, flow tills are characteristically very diverse in nature. The character of a unit of flow till consisting of several flow packages is summarised in Table 8.1. The clasts within flow tills show a broad range of characteristics, but are dominated by particles that are angular and have a non-spherical form. Clasts are commonly not striated or faceted. The deposit is usually dominated by sediment typical of high-level transport, but subglacially transported particles may be present. In general the size distribution is coarse and unimodal, although individual flow packages may be locally well sorted. Flow tills have a variable fabric, although individual flow packages can have strong fabrics, reflecting the former slope down which flow occurred. This type of sediment is poorly

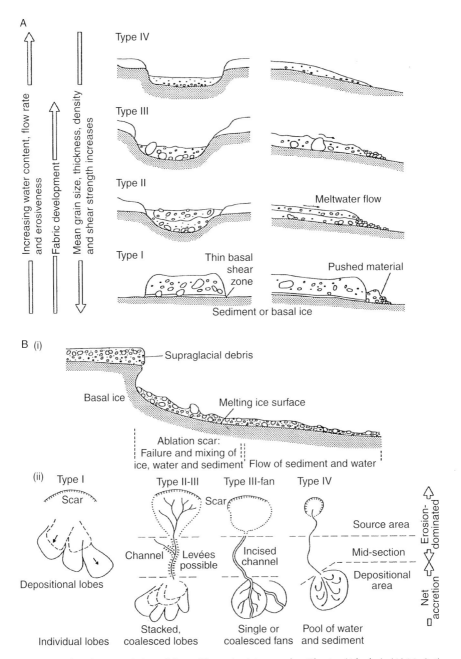

Figure 8.10   The characteristics of flow tills at the Matanuska Glacier (Alaska). (A) Variation of flow type with water content. (B) Morphology of the source and depositional area of different types of flow. [Modified from: Lawson (1982) *Journal of Geology* **90**, figures 3, 5 & 13, p. 282, 287 & 296]

consolidated with a low bulk density, although occasionally flow packages may be closely packed. Individual flow packages may sometimes be visible within a flow-till unit, and crude sorting within some flow packages may be present. The base of some flows may contain a concentration of larger clasts. Sorted sand and silt layers may be common, associated with reworking by meltwater on top of individual flows. Some flow packages may have erosional bases and small folds may also be present in certain flow types. In general they are highly variable and diverse sediments.

8.1.3 Distinguishing Different Types of Till

Distinguishing tills in the field can be challenging, particularly when dealing with ancient lithified tills, known as *tillites*, which record periods of ancient glaciation within the geological record (see Box 1.1). Deducing the process history of tillites is an important goal when studying such deposits (Table 8.1). It is worth considering a few of the key sedimentary properties and the information that they may contain, before discussing the broader issue of external facies relationships that are key to placing glacial sediments in their wider context.

The collection of clast-fabric data is one of the most traditional approaches to the description of tills. It has been used widely to infer information on ice flow, patterns of strain and as a basis for inferring till type, although its utility has been disputed by some (Box 8.2). The *clast fabric* of a till is defined by two properties: (i) the compass orientation of elongated particles; and (ii) the dip or angle at which those particles are inclined to the horizontal within the sediment. This information

---

BOX 8.2: TILL FABRIC: A GENETIC FINGERPRINT?

Till fabric analysis involves recording the compass orientation and dip of elongated clasts within a till. Generally only clasts with a pronounced long axis relative to the short and intermediate axes are analysed. Suitable clasts are carefully excavated from a cleared face of undisturbed till and the dip or inclination of each particle, along its long axis, is measured using a compass clinometer. The orientation of each particle, in the direction in which the long axis dips, is also recorded with a compass. The data are then plotted in a variety of ways – rose diagrams or equal area stereographic plot – to illustrate the preferred orientation (if present) and dip of the sampled clasts. The statistical distribution of clasts within a sample can be analysed in a variety of ways but it has become common place to do so via an analysis of eigenvectors. This multivariate method defines the mean orientation of the clasts within three-dimensional space and the degree of variability around this mean. It does this via three eigenvectors and three eigenvalues, which describe the degree of clustering around the three eigenvectors. Dowdeswell and Sharp (1986) showed how the fabric of different sediment from modern glaciers could be discriminated via plots of eigenvalues S1 and S2 (Diagram A), although Benn (1994) demonstrated how the data could be better illustrated via a ternary plot (Diagram B). This led to many authors attempting to use clast fabrics as a means of assigning a genesis to glacial diamicts. This approach was challenged by Bennett *et al.* (1999) who argued, on the basis of a large number of fabrics taken from sediments of known origin at modern glacier margins, that there is simply too much overlap between the fabric characteristics of different sediments to allow the genetic fingerprinting of diamicts on the basis of their clast fabric alone. Although not all workers agree with this view, it has led to a re-evaluation of the use of clast fabrics within glacial sedimentology. Although the fabric of glacial sediment can contain valuable information about the pattern or direction of cumulative strain, the key question is how to interpret this in terms of glacial processes.

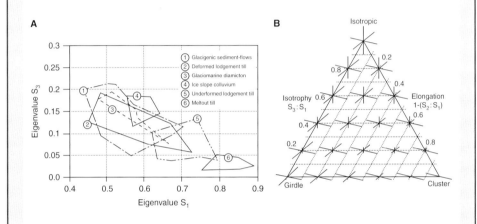

Sources: Bennett, M.R., Waller, R.I., Glasser, N.F., *et al.* (1999) Glacigenic clast fabrics: Genetic fingerprint or wishful thinking? *Journal of Quaternary Science*, **14**, 125–35. Dowdeswell, J.A. and Sharp, M.J. (1986). Characterization of pebble fabric in modern terrestrial glacigenic sediments. *Sedimentology*, **33**, 699–710. Benn, D.I. (1994). Fabric shape and interpretation of sedimentary fabric data. *Journal of Sedimentary Research*, **A64**, 910–5. [Modified from: Bennett *et al.* (1999) *Journal of Quaternary Science*, figure 1, p.126]

is normally portrayed on a rose diagram or on a stereograph. A variety of different statistical techniques have been used to describe fabrics, to determine the presence or absence of a preferred particle orientation, and to assess the distribution of particles about the preferred orientation (if present). Lodgement tills are tradition-ally thought to have a strong fabric in the direction of ice flow and the deviation or scatter of particles about the mean fabric is small. Similarly subglacial meltout and sublimation till should have a strong particle fabric parallel to the ice-flow direction but usually show a greater scatter of particles about the mean orientation. This reflects the fact that individual particles are disturbed as they meltout from the basal ice. Deformation till may possess well-developed particle fabrics orientated in the direction of tectonic transport, usually similar to the ice-flow direction. In contrast, supraglacial meltout and flow tills do not possess consistent particle fabrics, and particle orientation often varies through the deposit. In flow tills, for example, the fabric will reflect the direction of flow, which is normally downslope, and this will change over time as the ice topography changes during ablation. Similarly in supraglacial meltout tills, strong particle orientations may be recorded but these only reflect the orientation of former ice slopes against which the debris accumulated. Consequently, when sampled at several points through these deposits a random or scatter fabric is usually recorded.

Clast shape is also a helpful property in the interpretation of till origin (Table 8.1). As we saw in Chapter 7, debris transported without coming into contact with the glacier bed has very different particle shape and size characteristics from debris transported in contact with the bed. Subglacial tills should therefore contain a high proportion of subglacially transported debris with subglacial characteristics (i.e. rounded, striated, faceted and spherical clasts with a bimodal size distribution; Figure 7.9). In contrast supraglacial tills contain debris transported at high levels within the glacier and possess its characteristics (i.e. angular, non-spherical clasts with a coarse unimodal grain size distribution; Figure 7.6). These tills may also contain subglacial debris due to the upward transfer of basal debris by compression at the glacier margin, although in lower proportions. Particle size sorting within flow till may also be useful in their recognition. In very fluid flows, sediment sorting may occur and dewatering in other flows may give rise to pockets of sorted sediments and occasional layers of silts and sands. In addition, coarse boulder horizons may also be identified beneath some flow packages, formed as a tractive carpet of debris. Clast lithology may also be of some value in distinguishing till genesis, because subglacial tills tend to be dominated by local lithothogies, whereas supraglacial tills often contain a higher proportion of far-travelled lithologies transported on the surface of the glacier or englacially. Subglacial tills often contain evidence of shear, including features such as: clast smudges (brecciated clasts), low-angle shear planes and foliations, sole marks caused by erosion of the substrate and the extrusion of till into the underlying bedrock. In addition, glaciotectonites should contain evidence of folds and faults consistent with shear from overriding ice. In practice, however, there are very few hard and fast rules with which to interpret till genesis on the basis of internal sedimentary properties alone. This problem has become even more evident with the increasing use of micromorphology. Micromorphology involves the extraction of small *in situ* blocks of orientated sediment, which are then impregnated and thin-sectioned for study using a petrographic microscope. A variety of complex and sophisticated recording schemes now exist with which to document the observed structures and interpret these in terms of specific depositional and tectonic processes. This has provided valuable insight into the origin of many glacial deposits, but has also demonstrated that most glacial diamicts are formed by a combination of processes, thereby challenging traditional till classifications.

In many situations, the external setting or the context of a till unit is often more useful. In particular the association with surface landforms can be particularly diagnostic. For example, subglacial till is likely to be found in association with subglacial landforms such as drumlins and flutes (see section 9.2.1). In contrast, supraglacial tills such as flow or supraglacial meltout till are likely to be associated with areas of hummocky moraine (see section 9.1.3). Consequently the geomorphological context of the upper boundary or contact of a till unit may provide important insight into its origin. More recently the introduction of *facies analysis* has added further criteria. This approach is based not on the interpretation of individual units, but on the interpretation of the complete depositional sequence. It rests on the premise that most depositional environments can be characterised by distinctive associations or combinations of sediments or *facies* (*lithofacies*) and the bounding surfaces between them (Box 8.3). *Sedimentary facies* are bodies of sediment that are the product of a particular depositional environment or process and the relationship of one facies to another gives us the ability to assemble a picture of the depositional environment in which sedimentation occurred. At a modern glacier one deposit does not continue infinitely in any one direction, but will grade into other deposits. For example, a subglacial till surface may be dissected by a meltwater stream in which glaciofluvial sediments are being deposited. By studying the relationships between one deposit and the next we can reconstruct the depositional environment.

BOX 8.3: ARCHITECTURAL COMPONENTS OF A SUBGLACIAL TILL

Boyce and Eyles (2000) used an architectural element analysis to unpick the components present within a subglacial till (Northern Till) in Ontario, Canada. The method works by breaking down the individual components present (elements or lithosomes), while also focusing on the characteristics of the boundaries between these elements. The approach is to establish the individual building blocks that make up a complete facies and recognise a hierarchy of bounding surfaces that separate these building blocks. They used a range of different data sources to assist with this analysis, from detailed borehole logs and associated geophysics to the analysis of outcrop data. A range of architectural elements are identified within the till along with their bounding surfaces. The importance of this work is the recognition that subglacial tills are heterogeneous facies and that their sedimentation involves a range of subglacial processes that varies both in space and time. For example, the presence of boulder lines represent local episodes of erosion and reworking, and clastic dykes result from the local release of overpressured groundwater. Thin beds of sand and gravel may represent the location of subglacial drainage routes or short periods of bed–ice separation, and soft-sediment rafts result from the assimilation of sediment via deformation. The bounding surfaces between elements record the shifting pattern of erosion and deposition beneath the former ice sheet, as well as the lateral movement of different depositional processes through time. This level of detail is not always possible in the field and Boyce and Eyles (2000) used a superb three-dimensional data set, but the work illustrates the dynamic nature of subglacial sedimentation and the range of processes involved within it.

Source: Boyce, J.I. and Eyles, N. (2000). Architectural element analysis applied to glacial deposits: Internal geometry of a late Pleistocene till sheet, Ontario, Canada. *Geological Society of America Bulletin*, **112**, 98–118.

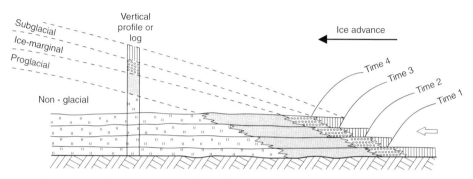

Figure 8.11 Illustration of Walther's law in the context of glacial environments.

To understand the relationship of one facies to another we must understand *Walther's law*. Walther studied recent sedimentary facies and their relationship to the environment in which they were deposited. From these observations he deduced that environments are not static through time and that as environments shift position, so the respective sedimentary facies of adjacent environments or processes succeed each other in a vertical profile (Figure 8.11). Therefore, in sequences where there is no apparent break in the sedimentary record the vertical profile of sedimentary facies is equivalent to the lateral variation of facies at any one time. In other words, if one was to turn a vertical profile on its side then it would give a picture of the lateral variation in the depositional environments present during the period of time represented by the vertical profile. In this way a vertical profile or section of glacial sediments can be translated into a picture of the particular glacial environment in which they were deposited. The formation of individual sediment units can then be explained. This is the principle by which *facies analysis* works. Vertical logs through sediment sections are constructed and the component facies described (Table 8.2); in this way the whole sequence is documented and then compared to modern depositional environments. Emphasis is therefore placed on interpreting the whole sequence of sediments and the environment in which they formed and not simply on the interpretation of specific units or components. The concept is illustrated with respect to a subglacial till in Figure 8.12.

The use of facies analysis provides a powerful field-based tool with which to interpret glacial sediments and the origin of till sequences. It does not, however, replace the need to examine the internal evidence present within

strong particle fabric. The process of lodgement may be interrupted locally by subglacial rivers, which rework the sediment into units of sand and gravel. These subglacial rivers are usually ephemeral and flow may switch on and off suddenly. The location of these rivers also varies through time. Changes in ice flow will also affect the continuity of the depositional processes and may cause erosional breaks. For example, different units of till, perhaps with different lithological clast contents, may be superimposed on top of one another. This occurs where different ice streams or ice lobes, from different source areas, compete with one another in a lowland area. As one lobe of ice waxes or wanes in strength its extent may vary with respect to other lobes, and therefore over time the lithological content of a till at any one point may vary depending on which lobe was dominated at that time. Sediment that is lodged is likely to experience subglacial deformation if the pore-water pressures are sufficiently high and the sediments may become overprinted by a range of tectonic characteristics. Figure 8.12 provides a generalised facies model for this type of environment.

In areas where a warm-based glacier moves over a soft substrate, subglacial deformation may occur as the underlying sediment is assimilated into the deforming layer. The thickness of the deforming layer will depend on the properties of the deforming sediment and on the shear stress applied to it by the glacier. As shown in Figure 8.4 low-levels of deformation are associated with simple overturning and folding, high-levels of deformation may involve intense folding and the development of tectonic laminations and boudins, whereas very high levels of deformation produce a homogeneous diamict. Sediment is transported within the deforming layer from areas of extending flow up to areas of compressional flow (Figure 8.6). Beneath areas of extending ice flow, deformation involves excavation and is defined by a sharp erosional surface along which new material is added to the deforming layer by downcutting of the deformation base. In contrast, in areas of compressive flow, the deforming layer thickens by the accumulation of till from up-ice areas. As a consequence, the deforming pile thickens by the addition of sediment at the top of the sequence and each successive tectonic state is preserved vertically in the sequence. At the base of the section there will be undeformed sediment, above which there will lightly folded and deformed sequences, followed by highly deformed sediment with tectonic laminations, boudins and other evidence of intense shear, and ultimately a homogenised diamict may occur at the top of the sequence if the level of deformation is sufficient. Figure 8.13 shows a model of the different deformation facies that might exist beneath an ice sheet. In practice, however, the sedimentary facies produced by subglacial deformation may be extremely complex depending on the deformation history and the character of the sediment that is being assimilated into the deforming layer.

2. **Deposition by cold-based glaciers.** Cold-based glaciers do not usually possess well-developed basal debris layers due to the absence of widespread glacial erosion and basal debris entrainment. Basal debris is derived in one of two ways: (i) by overriding of the frontal apron of fallen ice blocks and debris in front of the glacier snout; and (ii) freezing-on of water and debris draining from warm-based areas elsewhere in the ice sheet. Debris is usually frozen on in layers and highly folded and attenuated (Figure 8.14A). During glacier recession, the stratified basal-debris layers decay *in situ* and englacial debris is lowered onto

A: Constructional deformation (deposition)

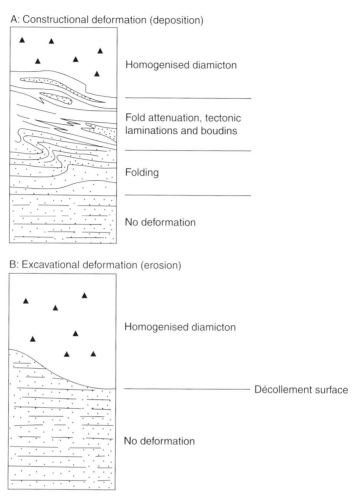

Homogenised diamicton

Fold attenuation, tectonic
laminations and boudins

Folding

No deformation

B: Excavational deformation (erosion)

Homogenised diamicton

Décollement surface

No deformation

Figure 8.13   Variation in subglacial glaciotectonic (deformation) facies beneath an ice sheet.
Constructional deformation occurs in areas of compression, whereas excavational deformation
is common under extending flow. [Modified from: Hart (1990) *Earth Surface Processes and
Landforms*, **15**, figure 3, p. 231]

the basal substrate. Interstitial ice is lost in cold arid areas by sublimation. Large
areas of sublimation or subglacial meltout till will result, which often retain the
crude stratification and folded structure of the englacial debris from which it is
derived. A typical vertical profile of the facies associated with cold-based glacier
is given in Figure 8.14B.

3. **Deposition by mixed-regime glaciers.** Many glaciers exhibit very thick basal
and englacial debris zones in response to repeated freezing-on of subglacial
meltwater draining from warm- to cold-based areas of the glacier and to thrusting at the warm–cold interface. A mixed thermal regime is, however, not the only
way in which this type of sediment assemblage may develop, because intense

BOX 8.4: THE INTERPRETATION OF MULTIPLE TILL SEQUENCES

Where thick sequences of supraglacial debris cover the glacier surface multiple layers of till – lodgement tills, meltout tills and flow tills – may be superimposed. In certain situations these may be separated by units of sand and gravel. A typical section consists of a till layer at the base, above which there is a sequence of sand and gravel, capped by a second till layer. This type of section is often referred to as a *tripartite till sequence*. Traditionally such sequences were interpreted as the product of multiple glaciations. The ice advanced to deposit the first till then retreated, depositing the sand and gravel, before readvancing to deposit the upper till. Many of these sequences have now been reinterpreted in terms of a single episode of glaciation. Ice advances over the area to deposit a basal lodgement till. As the ice retreats a topography of ice-cored ridges develops between which glaciofluvial sands and gravels are deposited. As the ice retreats further these outwash rivers become abandoned and the glaciofluvial deposits are covered by flow till derived from the adjacent ice-cored ridges. Meltout of the buried ice inverts the topography to give a tripartite till sequence. The correct interpretation of each till layer and of the facies present is therefore very important. One of the first sequences to be reinterpreted in this way was that at Glanllynnau in North Wales (Boulton, 1977). This represents an important step forward in the interpretation of till sequences and glacial stratigraphy.

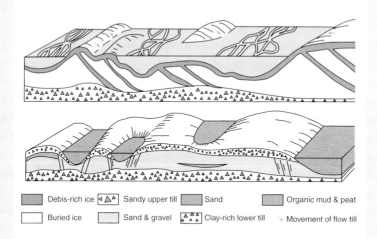

Debis-rich ice	Sandy upper till	Sand	Organic mud & peat
Buried ice	Sand & gravel	Clay-rich lower till	Movement of flow till

Source: Boulton, G.S. (1977) A multiple till sequence formed by a late Devensian Welsh ice cap: Glanllynnau, Gwynedd. *Cambria*, **4**, 10–31. [Modified from: Addison *et al.* (1990) *North Wales: Field Guide*. Quaternary Research Association, Cambridge, figure 17, p. 41.].

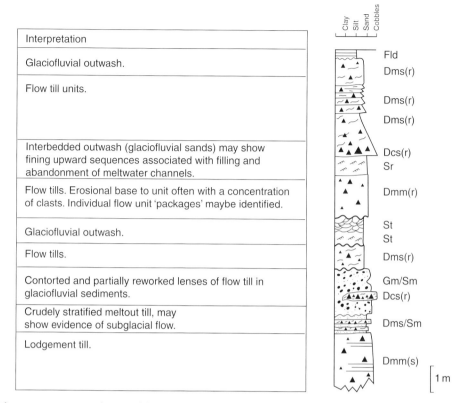

Figure 8.16   Typical vertical log and facies model for a glacier with a mixed thermal regime or high englacial debris content. [Modified from: Eyles (1983) *Glacial Geology*, Pergamon Press, Figure 1.8, p. 16]

This picture may be complicated for valley glaciers with thick supraglacial debris layers due to rockfall from adjacent valley sides. This supraglacial debris gives rise to a thick irregular drape of coarse, angular supraglacial meltout till, which is often reworked as flow tills and by fluvial action. This drape of debris transported at high levels within the glacier will mantle the subglacial sediment facies.

8.2 FLUVIAL SEDIMENTATION

Glacial meltwater entrains and transports sediment, which is subsequently deposited on, within, beneath or beyond the glacier. Sedimentation on or within the glacier may occur in surface channels and in either englacial or subglacial tunnels. Sedimentation in surface (supraglacial) channels is inhibited by the steep channel gradients and the smooth ice walls, which provide little frictional drag on the flow and its sediment load (Figure 4.1A). Deposition does, however, occur due to changes in the discharge of such channels, because sediment may lodge temporally in active channels as discharges fall. Sediment may also be deposited in channels where the gradient falls and the sediment load is high. Supraglacial channels are often ephemeral and when they are abandoned, sediment can be stranded within them.

Frictional drag is greater in subglacial tunnels that are in contact with the glacier bed. Sediments deposited within tunnels consist of sheet-like units of stratified sand and gravel in which secondary bedforms such as ripples, dunes and graded beds may occur. Deposition within such tunnels can be understood to some extent by the application of theory developed for the deposition of solids within pipes. Four flow regimes have been identified within pipes: (i) at low flow velocities a stationary bed with little or no transport occurs; (ii) as flow velocities increase material begins to slide over the bed as a single unit; (iii) at higher velocities suspension of all particles occurs, although the coarser fraction is still transported close to the tunnel floor; and (iv) at very high velocities all particles move in suspension and no size sorting is present. If the flow velocity was to fall rapidly during this final flow regime a massive, heterogeneous non-sorted sediment would result. Despite this work, very little is known about the processes of sedimentation within subglacial tunnels.

Subglacial meltwater may also deposit thin coatings of precipitated solutes in subglacial rock cavities (Figure 8.17). As we saw in Section 4.8.2 subglacial meltwater may dissolve soluble components and transport them. These may in turn be precipitated to form thin coatings. These coatings may either infill shallow depressions or be concentrated into small linear ridges. These precipitates are best developed on carbonate rocks such as limestone or chalk. In Norway, areas of bedrock that were once covered by basal cavities frequently contain a brown staining, which results from the precipitation of iron oxides from subglacial meltwater in the cavities.

Figure 8.17   Photograph of a subglacial calcium carbonate precipitate in front of the Glacier de Tsanfleuron, Switzerland. These precipitates are strung out parallel to the former ice flow, which was from right to left. [Photograph: M. Sharp]

The precipitation of silica has also been noted. In general these precipitates are confined to former subglacial cavities and are believed to be the product of regelation. As ice flows against a bedrock obstacle, pressure melting occurs on the upstream side and refreezing or regelation occurs on the downstream side (see Section 3.3.2). Refreezing concentrates the solutes within the meltwater, leading to their eventual precipitation in the lee of the bedrock obstacles. The linear form of many precipitates is due to smearing out of solute-rich meltwater by the flowing ice. These coatings are quickly dissolved and removed by weathering when the rock surfaces are exposed on deglaciation. Their long-term preservation potential is therefore small.

The processes of sedimentation beyond the glacier margin are much better understood. Sedimentation beyond the glacier occurs in the same way as conventional fluvial deposition, with the following exceptions.

1. The water is generally colder, denser and therefore more viscous. The viscosity of water increases with a fall in temperature. Increased viscosity reduces the settling rate for particles in suspension and allows a greater volume of suspended sediment to be transported.
2. The water and sediment discharge is highly seasonal. Water discharge beneath the Glacier d' Argentière varies, for example, between $0.1\,\mathrm{m}^3\,\mathrm{s}^{-1}$ in the winter and $11\,\mathrm{m}^3\,\mathrm{s}^{-1}$ in the summer. On Nisqually Glacier in North America the sediment transported during just 5 minutes in the month of June is equal to the whole sediment yield for the month of January. Similarly 60% of the annual sediment load from the Decade Glacier on Baffin Island was discharged during just 24 hours in 1965. Sediment discharge is therefore highly seasonal. It also varies diurnally. Most discharge occurs during the period of nival floods early in the melt season (see Section 4.7).

Sediment is transported both in suspension and as bedload (traction and saltation). A number of studies have attempted to record the sediment load of meltwater streams and its variation through time. Suspended sediment can be measured relatively easily by water sampling; the water sample is then filtered or evaporated to determine the sediment content. Results show that during the winter, when discharge is negligible, meltwater contains only a few milligrams of sediment per litre. During summer, this rises to several grams per litre. Suspended sediment content reaches a peak early in the summer as the fluvial system within the glacier is flushed clean (see Section 4.8.2). Suspended sediment content also varies diurnally and peaks prior to the maximum daily discharge on many glaciers (Figure 4.10). In contrast accurate estimates of the sediment moving as bedload are much more difficult to obtain and seasonal variations are less well understood at present (Box 8.5). Estimates of the relative importance of the two components (suspended versus bedload) vary from as little as 40% suspended load to over 90% of the total sediment discharge, depending on the particular characteristics of the glacier. Sedimentation in front of a glacier can be divided into three zones, although the boundaries between each are somewhat unclear. The three zones are: (i) the proximal zone; (ii) the medial zone; and (iii) the distal zone.

BOX 8.5: SEDIMENT TRANSPORT IN GLACIAL MELTWATER STREAMS

Reliable estimates of bedload transport within meltwater streams are difficult to obtain, due to the high flow magnitudes and large sediment volume involved. One of the first studies to provide reliable estimates of bedload transport was that of Østrem (1975). Two techniques were used to measure the bedload of a proglacial stream in front of the glacier Nigardsbreen in Norway.

The first method involved the construction of a 50 m steel fence across the main meltwater stream during the summer of 1969. The mesh size was such that it trapped all sediment larger than 20 mm in diameter. The accumulation of coarse bedload trapped by the fence was measured by probing the depth twice a day at 176 points along the fence.

The fence survived for 3 weeks before being destroyed in a flood. Between the 24 May and 19 June, 400 tonnes of material were trapped. Samples of meltwater were also taken during this period to determine the suspended sediment content. During this period approximately 1200 tonnes of suspended sediment were discharged. This suggests that during the study period bedload transport accounted for 25% of all the material transported, although it must be noted that sediment in the size range of 1–20 mm in diameter was not trapped and therefore this value must be regarded as a low estimate.

The second method involved the annual survey of a delta formed as the meltwater from Nigardsbreen enters a lake about 1 km in front of the glacier. Most, if not all, of the bedload in transport is deposited on this delta. The average annual accumulation on this delta is 11 200 tonnes, giving a crude estimate of the total bedload moved each year by the meltwater streams.

In recent years the sophistication of bedload traps has improved and accurate estimates of bedload transport are more common place. However, recent work simply confirms the high levels of bedload transport within meltwater streams first quantified in detail by Østrem.

Source: Østrem, G. (1975) Sediment transport in glacial meltwater streams, in *Glaciofluvial and Glaciolacustrine Sedimentation* (eds A.V. Jopling and B.C. McDonald), The Society of Economic Paleontologists and Mineralogists, Special Publication 23, pp. 101–22.

1. **The proximal zone.** In this zone sedimentation is dominated by: (i) the changing position and geometry of the ice margin and/or any ice-cored ridges present (Figure 8.15); (ii) the rate of supply or availability of supraglacial meltout till; and (iii) the seasonal flood regime. Braided stream flow is only one part of the total hydraulic system. Resedimentation of supraglacial meltout till as mud, debris and subaqueous flows is common and may dominate the depositional processes. The availability of large quantities of readily transported sediment and the rapid build up and decay to and from flood discharges has a strong effect on the

character of the fluvial sediments deposited. Meltwater streams are often incompetent to transport the available till load and may simply redistribute it as structureless, matrix-supported, outwash unit in which the particle-size distribution is little different from the parent till. During flood phases all available particle sizes may be transported and deposited simultaneously. Weak stratification at the top of these massive units of outwash may develop during the waning flood stage and consists of individual lamellae or sediment layers several clasts thick formed by the removal of the fines (*winnowing*) to give an armoured, often scoured, surface. Proximal outwash is frequently found interbedded with units of flow till and other mass-flow deposits, particularly where melt streams are bordered by ice-cored debris ridges. Consequently, the massive unstratified proximal outwash typical of this zone is difficult to distinguish from supraglacial meltout till.

Deposition also frequently occurs on buried ice, the meltout of which causes deformation or subsidence structures. These usually consist of normal or extensional faults and synclinal fold or sag structures (Figure 8.18). If meltout and subsidence occurs while deposition is still taking place these subsidence structures are referred to as *syn-sedimentary*. Evidence for subsidence may not be present on the surface, because meltwater deposition is concentrated in the subsiding areas and infills them (Figure 8.19). If meltout occurs after deposition has finished then the landsurface is deformed and parallels the bedding (Figure 8.19). These subsidence structures are *post-sedimentary* (see Section 9.3).

Figure 8.18 Faulted sands within outwash sediments in front of Skeidararjökull, Iceland. These normal faults formed during subsidence associated with the meltout of buried ice. [Photograph: M. R. Bennett]

A Syn-sedimentary subsidence

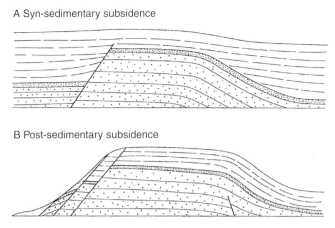

B Post-sedimentary subsidence

Figure 8.19   Syn-sedimentary and post-sedimentary subsidence structures

2. **The medial zone.** Away from the ice margin a braided river pattern develops, in
   which ephemeral *bars* and channels dominate (Figure 8.20). Individual channels
   and bars vary in size from a few metres to hundreds of metres in width. The
   depth of channels is typically only a few metres. The braided pattern develops in
   response to the large sediment load, high discharge and the steep gradient of
   outwash surfaces. The particle size of sediment deposited rapidly falls away
   from the glacier margin, which reflects the decline in stream power and the

Figure 8.20   Braided outwash plain, Entujökull, Iceland. [Photograph: M. R. Bennett]

processes of clast attrition in highly turbulent channels (Figure 8.21). Three types of bar or sandwave can be identified, although they are not unique to glacial outwash systems (Figure 8.22).

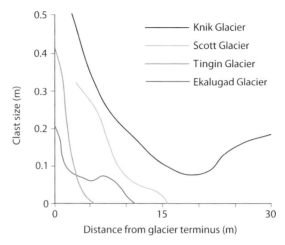

Figure 8.21   Relationship between clast size and distance from the glacier margin for a variety of glacial outwash systems. [Modified from: Drewry (1986) *Glacial Geologic Processes*, Arnold, figure 10.10, p. 159]

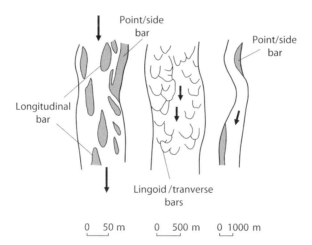

Figure 8.22   Schematic illustration of the three main types of bars found in braided outwash channels. [Modified from: Drewry (1986) *Glacial Geologic Processes*, Arnold, figure 10.11, p. 160]

A. The *longitudinal bar*. This forms in a mid-channel position when the coarsest part of the stream load is deposited as stream flow falls and loses competency. These bars are orientated roughly parallel to the current flow. They are small when first formed, but continue to grow in length and height as fine sediment is deposited downstream in the lee of the bar. Grain size tends to decrease downstream.

B. The *linguoid* or *transverse bar*. This type of bar is orientated transverse to the direction of stream flow and may possess a diamond or rhombic shape with a steep downsteam face. They develop under high flow conditions and extend by particles avalanching down their front face.

C. The *lateral*, *side* or *point bar*. These bars are typically very large and develop on the sides of stream channels in quieter water. They are attached to the sides of the channel.

With distance from the glacier margin the proportion of linguoid or transverse bars increases as does the sand component within the fluvial system. Sediments resulting from this braided channel system range from boulders to sands. The sedimentary structure within these sediments reflects five processes that operate in the braided channel system. These are: (i) the formation of bars; (ii) the formation of bedforms such as ripples and dunes in finer material; (iii) the erosion (scour) and fill of channels; (iv) the deposition of finer sediments during low flows, particularly in backwaters; and (v) overbank sedimentation during the falling stage of flood flows. Table 8.3 shows the types of sedimentary facies associated with these processes.

Table 8.3   Diagnostic criteria for recognition of common glaciofluvial stream deposits.

Code	Facies	Interpretation
Gm	Massive, matrix-supported gravel with no sedimentary structure	Debris flow
Gms	Massive or crudely bedded gravel with horizontal bedding and clast imbrication	Longitudinal bars and channel lag deposits
Gt	Stratified gravel with trough cross-beds	Minor channel fills
Gp	Stratified gravel, planar cross-beds	Transverse or linguoid bars
St	Medium to very coarse pebbly sand with solitary or grouped trough cross-beds	Dunes
Sp	Fine to very coarse pebbly sand with solitary or grouped planar cross-beds	Transverse or linguoid bars
Sr	Very fine to coarse sand, with ripple marks	Ripples, low-flow regime
Sh	Fine to very coarse often pebbly sand, with horizontal laminations	Planar bed flow, high-flow regime
Ss	Fine to coarse sand, may be pebbly, with broad shallow scour structures	Minor channels or scour hollows
Fl	Sand, silt and mud, with ripple marks	Waning flood deposits and overbank deposits
Fm	Mud and silt with desiccation cracks	Drape deposits formed in pools of standing water

For a detailed description of the types of cross-bedding referred to in the table the reader is referred to the volume by: Collinson, J.D. and Thompson, D.B. (1989) *Sedimentary Structures*. Unwin Hyman, London.

[Modified from: Eyles et al. (1983) *Sedimentology*, **30**, table 1, p. 395]

3. **The distal zone.** The proportion of fine-grained sediment increases dramatically with distance from the glacier margin. During normal discharge the main flow is concentrated in a single channel, although at peak discharge a braided pattern may develop. The glacial influence decreases and the flow regime is dominated less by the seasonal patterns of ice melt. There is a gradual transition into a more conventional fluvial system.

The pattern of sedimentation in the proximal, medial and distal zones will be disrupted if the glacier is subject to catastrophic floods or jökulhlaups. Sedimentation will be partly controlled by the shape of the flood hydrograph, which is a function in part of the cause of the jökulhlaup (see Figure 4.12). Floods triggered by volcanic activity beneath a glacier are typically of high magnitude but of short duration, in contrast to floods caused by the drainage of ice-dammed lakes, which tend to be of longer duration. Research to date suggests that jökulhlaup discharges result in sediment sequences consisting of massive, poorly sorted, non-graded or inversely graded sediments. The latter are characterised by large surface boulders, around which flow may be channelled. If present, these channels tend to contain boulder lags, fields of mega-ripples and streamlined boulder hummocks. The main unit is interpreted as the product of hyperconcentrated fluid–sediment mixtures that are sufficiently dense to transport large boulders on their surface and prevent any size-sorting or grading from developing. These hyperconcentrated flows are associated with the initial flood surge. As the flow stage declines the sediment solidifies and dewaters. This water may then form a series of more fluid flows on the surface of the main deposit. If these fluid flows are sufficiently large they may scour channels and deposit the lag horizons. The massive units of boulder-rich gravel typical of jökulhlaups provides a sharp contrast with the deposits produced by normal glacial discharges.

8.3 SUMMARY

Glaciers on land deposit sediment either directly or via meltwater. The product of direct glacial sedimentation is a glacial till. A till is defined as a sediment with components that are brought into contact by the direct agency of glacier ice and which has not been disaggregated, although it may suffer glacially induced flow either in the subglacial or supraglacial environment. Traditional classifications of glacial till recognise six main types – lodgement till, subglacial meltout till, deformation till, supraglacial meltout or moraine till, flow tills and sublimation till. Recent work suggests that many of these processes occur together and that it is only possible to recognise subglacial and supraglacial tills. The properties of these two end-members are more distinct and partly depend on the transport pathway followed by the debris within the glacier. Glacial sedimentary facies are primarily controlled by basal thermal regime. Deposition by meltwater occurs either within the glacier or beyond its margins. The type of sediment and sedimentary facies produced by glacial meltwater is primarily controlled by the distance from the ice margin. Close to the ice front the sediments are coarse, poorly sorted and often chaotic. With distance from

Source: Larsen, E., Kjaer, K.H., Jensen, M., *et al.* (2006) Early Weichselian palaeoenvir-
onments reconstructed from a mega-scale thrust-fault complex, Kanin Peninsula,
northwestern Russia. *Boreas*, **35**, 476–92. [Modified from: Larsen *et al.* (2006) *Boreas*,
35, figure 5, p. 489]

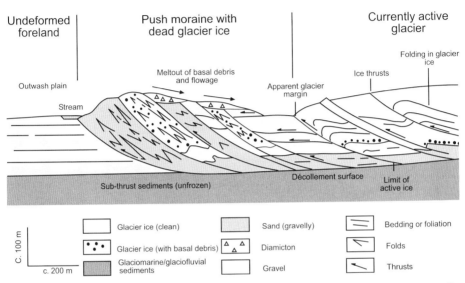

Figure 9.5 Schematic cross-section through the Uvêrsbreen push moraine in Svalbard.
[Modified from: Hambrey and Huddart (1995) *Journal of Quaternary Science*, **10**,
figure 11, p. 324]

the glacier snout terminated (Figure 9.6). A moraine system created by subglacial deformation has also been described from a surge of Brúarjökull in
Iceland. Here variation in the geometry of the moraine can be attributed to the
hydrogeological properties of the foreland (Box 9.2).

Large glaciotectonic moraine systems may also incorporate bedrock rafts. One of
the most famous examples is Møns Klint in Denmark, where large thrust-blocks of
Cretaceous Chalk occur in tectonic slices with glacial sediment forming a series of
ridges, the structures of which are revealed in spectacular coastal cliff sections.
Similar, although older, chalk rafts are exposed in coastal cliffs at Sidestrand
Norfolk, England and other examples of systems incorporating rock rafts include
the Dirt Hills in Saskatchewan, Canada. The Dirt Hills have been pushed up over
300 m and have been partly overridden by an ice sheet during the last glacial cycle.
They consist of hundreds of parallel ridges, transverse to the former ice flow,
between which there are small linear lakes. The ridges comprise blocks of sandstone bedrock that have been thrust forward over a basal décollement surface
located in softer clay-rich beds beneath. It has been suggested that rapid glacial

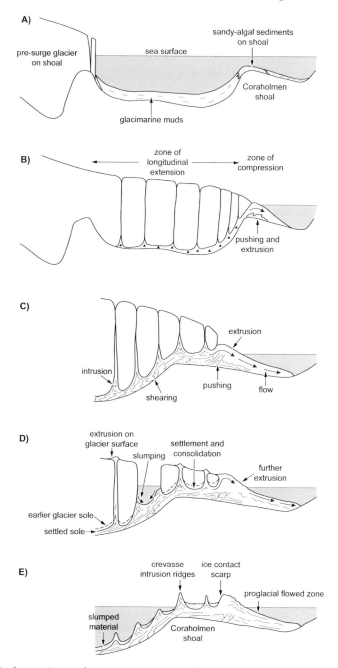

Figure 9.6 Push moraine and crevasse-squeeze ridges formed by a surge of Sefströmbreen in Svalbard. This moraine system appears to be the product of subglacial deformation as the ice advanced across glaciomarine and marine muds on a fjord floor before mounting and terminating on the island of Coraholmen. [Modified from: Boulton *et al.* (1996) *Quaternary Science Reviews*, **15**, figure 22, p. 144]

BOX 9.2: SUBGLACIAL DECOUPLING AT SEDIMENT–BEDROCK INTERFACE AND MORAINE FORMATION

Kjær *et al.* (2006) describe a series of observations from Brúarjökull in Iceland, a northern outlet glacier of the Vatnajökull Ice Cap in Iceland. The glacier has a flow regime which oscillates between major winter flow events that last up to 3 months, punctuated by quiescent periods that last between 70 and 90 years; surges occurred in 1890 and 1963 and involved ice-marginal advances of between 10 and 8 km in each case. Since 1964 ice flow has been negligible and the ice margin retreated at a rate of up to 250 m per year between 2003 and 2005, with a surface lowering of 6–7 m per year. Kjær *et al.* (2006) argue that fast ice flow during the 1890 surge occurred because of decoupling between the deforming bed and the bedrock as evidenced by thin layers of waterlain sediments along this interface. Subglacial drainage was unable to escape through the impermeable basalt bedrock and consequently pore-water pressures built up at the sediment–rock interface, leading to rapid deformation and or sliding. The limit of the glacier advance is determined by locations at which this pressurised groundwater could escape.

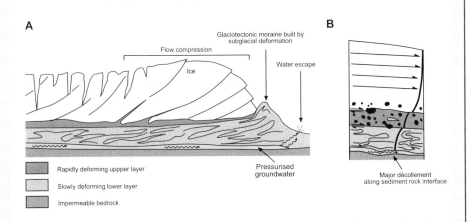

This limit is associated with a large glaciotectonic moraine, which is described by Benediktsson *et al.* (2008). As a result of substrate–bedrock decoupling during the surge, subglacial sediment was advected over bedrock and deformed compressively, leading to a gradual thickening toward the ice margin and the formation of a sediment wedge in the marginal zone. Water escape in these areas at the end of the surge led to a substrate–bedrock coupling and stress transfer into the sediment sequence below the ice margin, causing brittle deformation in areas of sand and gravel and more ductile deformation in areas of finer-grained sediments. The glacier increasingly ploughed into these sediments, leading to marginal moraine formation; narrow ridges occur where fine-grained incompetent sediments crop out and were deformed in a ductile fashion, whereas in areas of coarse-grained sediment deformation occurred via brittle failure along thrusts to give a wider and topographically more complex moraine. Rapid emplacement of the moraine is envisaged, taking just a few days towards the end of the surge. Moraine formation occurred therefore as a two-stage process, first the sediment wedge was built up by sediment transport within the subglacial deformating layer, before the ice-marginal sediment pile was tectonised further as the ice bed coupled with the deformating layer more effectively once it ceased to move freely over the underlying bedrock upon dewatering.

Sources: Kjær, K.H., Larsen, E., van der Meer, J., *et al.* (2006) Subglacial decoupling at the sediment/bedrock interface: A new mechanism for rapid flowing ice. *Quaternary Science Reviews*, **25**, 2704–12. Benediktsson, Í.Ö., Möller, P., Ingólfsson, Ó. *et al.* (2008) Instantaneous end moraine and sediment wedge formation during the 1890 glacier surge of Brúarjökull, Iceland. *Quaternary Science Reviews*, **27**, 209–34. [Reproduced from: Benediktsson *et al.* (2008) *Quaternary Science Reviews*, **27**, 3figure 14, p. 232. Copyright © 2008, Elsevier Ltd].

loading of the competent sandstone bedrock and the saturated incompetent mudstone and clay-rich strata beneath caused the sandstone to fracture and be thrust-up in a complex series of blocks along the margins of the advancing ice sheet.

The above review illustrates the range of different morphologies and scales at which glaciotectonic moraines are encountered at past and present ice margins. It is worth exploring further some of the variables that control the development of these moraines. Figure 9.7 defines some of the key variables, which include the following.

1. **Application of stress.** A glacier can transfer stress to the foreland in several different ways, each having a different impact. At its simplest a glacier can be viewed as a bulldozer pushing from the rear. However, in practice it may also apply stress via gravity-spreading. A glacier increases in thickness away from the ice margin and consequently the load applied to subglacial sediments also increases away from the margin. This effectively sets up a lateral stress gradient from areas of thick ice and high load to areas at the margin of thin ice and little load. This may be as important in driving proglacial deformation as the force applied directly by the glacier.

2. **The geometry of the foreland wedge.** This can be defined in terms of the size, both the proximal to distal width and depth of the proglacial area (foreland) that was, or is, being tectonised to produce the moraine. In the case of a small seasonal push moraine it is likely to consist simply of a thin veneer of surface debris, alternatively it might involve a significant part of a glacier proglacial area or perhaps part of the ice margin itself. Prior to deformation the foreland wedge can be defined as a slab of known width (measured

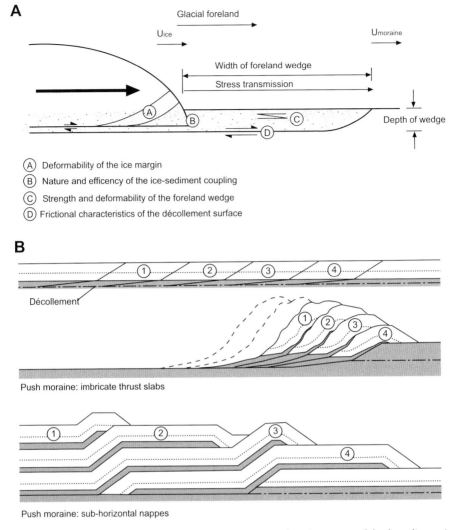

Figure 9.7 (A) The anatomy of a glaciotectonic moraine showing some of the key dimensions and properties involved in their formation. (B) Alternative tectonic models for the formation of glaciotectonic moraines.

perpendicular to the ice margin) and depth. Many large push moraines appear to involve slabs with a high aspect ratio (i.e. slabs that are thin, but laterally extensive). The topographic geometry of the foreland may also be important. For example, if a glacier is advancing up a reverse slope it will transfer the forward stress effectively into the slope, pushing up a larger moraine; a glacier on a horizontal surface will not have the same impact because much of the forward stress is not transferred into the proglacial sediment (Figure 9.8). An obvious nucleus against which an ice

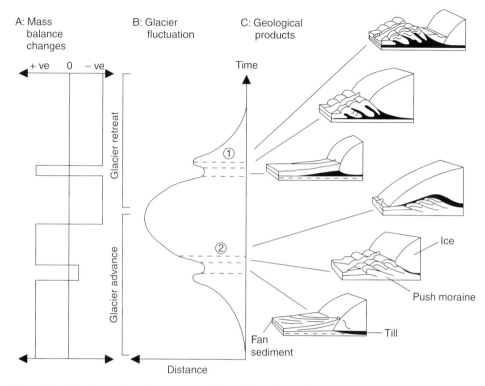

Figure 9.8   Push-moraine-forming episodes within a glacial cycle. (1) Glacier decay is halted by a period of positive mass balance and an outwash fan forms, which is subsequently pushed up by a glacier advance due to the continued positive mass balance. As a negative mass balance re-establishes itself and ice retreat continues, the glacier decays away from the moraine and meltwater is trapped between it and the moraine to form kame terraces. (2) A period of glacier advance is halted by a short spell of negative mass balance. An outwash fan first forms at the stationary ice margin before being pushed up into a push moraine as a positive mass balance regime re-establishes itself and the glacier advances. The implication is that during a glacier advance push moraine formation is initiated by a climatic amelioration, whereas during decay it is stimulated by a deterioration in climate. [Modified from: Boulton (1986) *Sedimentology*, **33**, figure 17, p. 695]

margin can push is an outwash fan and the association of glaciotectonic moraines with the former location of outwash fans has been widely noted. This observation has been used to build a general model of where push moraines may occur on a glacial advance–retreat cycle (Figure 9.8). Consider a glacier experiencing a period of long-term advance, due to a positive mass balance, where continuous frontal advance would plough up only a small push moraine given a horizontal glacier forefield. Its size will remain limited by the loss of material beneath the advancing glacier and it will grow only where there is an upstanding or resistant sediment mass against which the glacier may push effectively. If, however, the ice-marginal

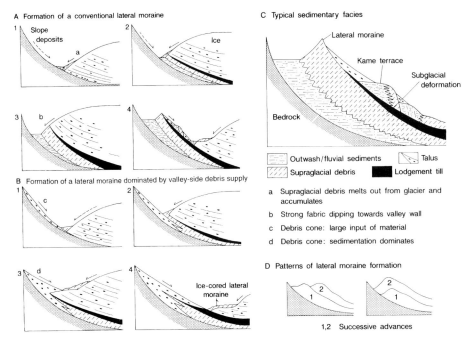

Figure 9.11 Morphology and sedimentary facies of lateral moraines. (A) Formation, morphology and facies of a lateral moraine where the supply of supraglacial debris is greater than the supply of valley-side debris. (B) Formation, morphology and facies of a lateral moraine where the supply of supraglacial debris is less than the supply of valley-side debris. (C) Typical sedimentary facies of a lateral moraine. (D) Possible modes of growth for a lateral moraine formed by successive ice advances.

modified by ice-marginal meltwater, which can deposit material between the lateral moraine and the valley wall. Small kame terraces (see Section 9.3) may be superimposed upon the proximal slopes of lateral moraines as the ice retreats (Figure 9.11). Lateral moraines often display cross-glacier asymmetry; the lateral moraine on one side of a valley being larger than that on the other. This asymmetry reflects the distribution of debris within the glacier, which is normally a function of the distribution of bare rock faces in the upper reaches of the glacier.

Ice-marginal fans composed predominately of diamicts represent one type of dump moraine. Here a stationary ice margin subject to seasonal or more prolonged stillstands becomes the focus for the build up of an ice-marginal fan consisting of a diverse range of materials delivered by a range of different processes. Sediment may be delivered by debris flows originating from supraglacial sediment, by direct avalanching of debris, or by small seasonal melt-streams, building up a layered internal stratigraphy. The pile of debris may become a focus for deformation and ice pushing, but it essentially accumulates through a diverse range of processes transferring sediment in an off-ice direction. These grade with increasing glaciofluvial input into outwash fans (see Section 9.3). These fans have been used to explain one component of the landform assemblage commonly found in the hills of upland Britain and associated with the Younger Dryas (Box 9.4).

BOX 9.4: INTERPRETING LANDFORM ASSEMBLAGES: EXAMPLES FROM THE SCOTTISH HIGHLANDS

The mountains of the British Isles were last occupied by glacier ice during a short sharp return to cold conditions at the end of the last glacial cycle, which is known as the Younger Dryas. Glaciers left a distinctive topography of mounds and ridges that became known as Scottish Hummocky Moraine. These deposits were interpreted in the late 1960s and 1970s as the product of ice stagnation; a theory consistent with emerging evidence of the rapidity of climate warming after the Younger Dryas. This model became the dominant paradigm and the geometry of these former ice bodies were mapped out on the basis of the extent of such deposits and palaeoclimatic inferences made on the basis of these palaeoglaciological reconstructions. This non-genetic landform interpretation was challenged in the late 1980s and early 1990s by several different glacial geologists who started to see Scottish Hummocky Moraine as an assemblage of ice-marginal landforms that could be deciphered to reveal a picture of active rather than passive deglaciation (Bennett, 1994). Since then the focus has been on interpreting this landform assemblage and deciphering its palaeoglaciological significance. In recent years these models have become increasingly sophisticated demonstrating the glaciological potential of Britain's upland areas (Benn and Lukas, 2006). The case study demonstrates how paradigms shift though time and the importance of the objective appraisal of field evidence.

Sources: Benn, D.I. and Lukas, S. (2006) Younger Dryas glacial landsystems in north west Scotland: an assessment of modern analogues and palaeoclimatic implications. *Quaternary Science Reviews*, **25**, 2390–408. Bennett, M.R. (1994) Morphological evidence as a guide to deglaciation following the Loch Lomond Readvance: a review of research approaches and models. *Scottish Geographical Magazine*, **110**, 24–32.

9.1.3 Ablation Moraines

Material on the surface of a glacier may become concentrated at the ice margin. This results from the supply of supraglacially transported debris and the transfer of subglacial and englacial material to the ice surface by upward flowing ice and englacial thrust planes (see Section 7.4). As supraglacial debris accumulates it initially accelerates ice melt, because darker surfaces absorb solar radiation more effectively than light surfaces, which are more reflective. However, surface melting is retarded as debris thickness increases because it insulates the ice from surface heating. If the debris cover is sufficiently thick, the glacier margin may become detached from the main body of the glacier and become stagnant, resulting in an ice-cored moraine or ablation moraine. The morphology of ablation moraines is controlled by the distribution of debris on and within the glacier, which may be either a product of the glacier structure – thrusts and folds within it – or the result of the surface distribution of supraglacial debris (see Section 7.1). The supraglacial debris distribution is controlled by the location of medial moraines on the glacier and the presence of patches of rock-fall debris.

The type of landform produced depends on two main variables: (i) the debris concentration; and (ii) the nature of the debris supply (i.e. continuous or discontinuous). If the debris concentration on the glacier surface is low and it is concentrated along the ice margin, then discrete moraines will tend to form by dumping and ablation of a narrow belt of buried ice along the ice margin (Figure 9.12). However, if the debris concentration is high and it is spread over a larger part of the glacier an area of *hummocky moraine* will result (Figures 9.12 and 9.13). Hummocky

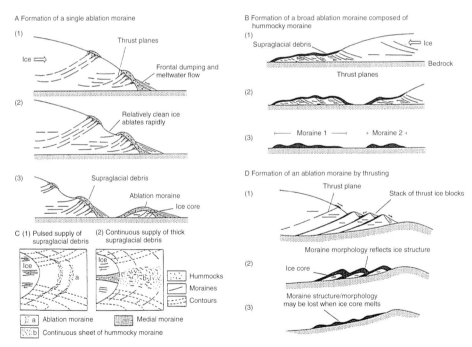

Figure 9.12   The morphology and formation of ablation moraines. (A) The formation of simple ablation moraines by differential ablation of the ice surface. (B) The formation of a complex ablation moraine consisting of belts of hummocks moraine. (C) The distinction between ablation moraines formed from discontinuous and continuous rates of supply of supraglacial debris. (D) Formation of an ablation moraine by thrusting.

Figure 9.13   Hummocky moraine in front of Ossian Sarsfjellet, Svalbard [Photograph: N.F. Glasser]

moraine –an irregular collection of mounds and enclosed hollows – is formed by the meltout of ice-cored debris. It may form a continuous sheet of irregular hummocks if the debris on the ice surface is continuous or alternatively discrete belts of hummocky moraine may result if the debris on the glacier surface is not continuous or of a constant thickness (Figure 9.12). This situation can arise if the supply of debris is either pulsed, due for example to discontinuous episodes of rockfall activity, or controlled by the spatial pattern of thrusts within the glacier (Figure 9.12). The morphology of these moraines may reflect the englacial structure of the glacier, at least while they retain their ice core (Figure 9.12). This morphology is often lost, however, when the ice core ablates and the size of the moraine is radically reduced. It is important to note that hummocky moraine can form by other mechanisms as well (see Section 9.2.1).

The processes responsible for the de-icing of ice-cored moraines have been monitored at Kötlujökull over a number of years (Figure 9.14). Here melting along the bottom surface of the ice cores occurs at an annual rate of 250 mm and involves a complex array of processes, including the development of sinkholes at the toe of dead-ice blocks in response to local melting. These initiate retrogressive rotational sliding or backslumping of the ice-cored slopes, and the formation of distinct edges and areas of exposed ice accelerate the rate of melting. Topographic inversion occurs in the initial, but not later, phases of de-icing, where general lowering of the surface occurs.

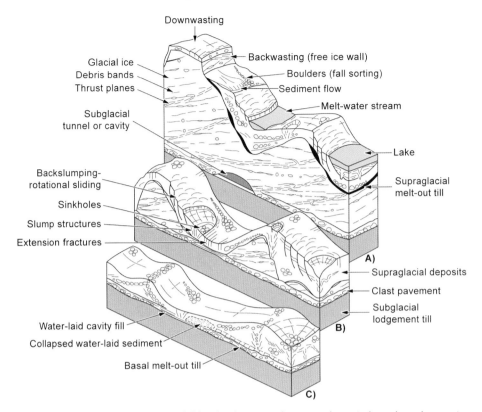

Figure 9.14   Sedimentological model for the de-icing of ice-cored terrain based on observations at Kötlujökull, Iceland. [Reproduced with permission of Blackwell Publishing Ltd from: Kjær and Kruger (2001) *Sedimentology*, **48**, figure 12, p. 948]

9.2 SUBGLACIAL LANDFORMS FORMED BY ICE OR SEDIMENT FLOW

This category of landforms is divided into those which have been ice-moulded and those that have not. Ice-moulded landforms are significant because they provide information about the direction and velocity of glacier flow.

9.2.1 Ice-Moulded Subglacial Landforms

Three broad families of ice-moulded subglacial landforms (bedforms) have been identified on the basis of size (Figure 9.15). Although each may be genetically distinct, these are as follows.

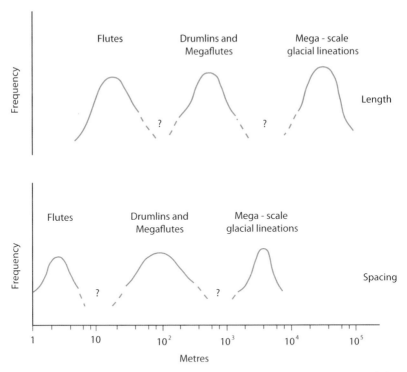

Figure 9.15 Schematic representation of principal spatial frequencies and lengths of streamlined subglacial landforms. The data suggest that there are three populations of subglacial bedform: (i) flutes; (ii) drumlins and megaflutes; and (iii) megascale glacial lineations. [Reproduced with permission of John Wiley & Sons Ltd from: Clark (1993) *Earth Surface Processes and Landforms*, **18**, figure 6, p. 9]

1. **Flutes.** Typically these are low (< 3 m), narrow (< 3 m), regularly spaced ridges which are usually less than 100 m long and are aligned parallel to the direction of ice flow (Figure 7.8B). They have a uniform cross-section and usually start from either: (i) a large boulder; (ii) a collection of boulders; or (iii) a bedrock obstacle. They are typically composed of lodgement till, although they may also contain fluvial sands and gravels. Clusters of boulders may occur within the body of the flute. They are a common landform in front of many glaciers today.

2. **Drumlins, megaflutes and rogen (ribbed) moraines.** Drumlins are typically smooth, oval-shaped or elliptical hills composed of glacial sediment (Figures 9.16 and 9.17). They are between 5 and 50 m high and 10–3000 m long. They have length to width ratios that are less than 50. They are composed of a variety of materials, including: (i) lodgement till; (ii) bedrock; (iii) deformed mixtures of till, sand and gravel; and (iv) undeformed beds of sand and gravel. They tend to occur in distinct fields or drumlin 'swarms' and are not uniformly distributed beneath a glacier. The formation of drumlins by small glacial read-vances suggest that drumlins may form rapidly. In contrast to flutes, Megaflutes

Figure 9.16 Oblique aerial photograph of a drumlin swarm in Langstrathdale northern England. Ice flow was from left to right. [Photograph reproduced with permission from: Cambridge University Collection of Aerial Photographs]

are taller ($<$ 5 m), broader and longer ($>$ 100 m) and are distinguished from drumlins by having a length to width ratio in excess of 50. Their long axis is parallel to the direction of basal ice flow and they typically have a uniform cross-section. Occasionally they may start from a large bedrock obstacle, but most do not. Rogen (ribbed) moraines have a variety of morphological forms but generally consist of ridges formed transverse to flow, although sometimes they show reshaping parallel to ice flow.

3. **Megascale glacial lineations (MSGL).** In recent years, much larger (megascale) lineations composed of glacial sediment have been recognised on satellite images. They are typically between 8 and 70 km long and between 200 and 1300 m wide, with 300–5000 m spacing between lineations. On the ground their morphology is often difficult to detect.

Subglacial bedforms may occur superimposed one on top of another, for example with smaller bedforms resting on the backs of larger ones. Megaflutes may be superimposed on the backs of larger drumlins, and in many of the drumlin fields of northern England small drumlins are found on larger drumlins. The orientation of the two sets of bedforms can either be coincident or discordant. In the latter case the bedform patterns provide evidence of changing patterns of ice flow (see

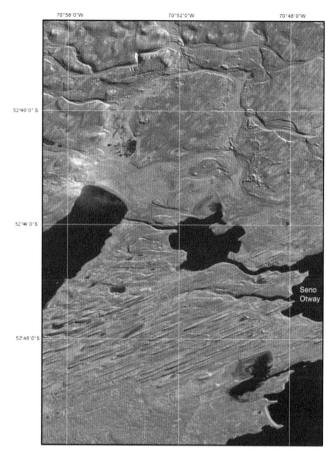

Figure 9.17   Glacial lineations and drumlins on a Landsat satellite image of Patagonia, southern South America. North is at the top of the image, and former ice flow was from SSW to NNE. The image is approximately 50 km wide. [Image courtesy of: N.F. Glasser]

Chapter 12). In a hypothetical ice sheet, ice will flow from the ice divide to the margin and this pattern will be recorded by subglacial bedforms. Within this ice sheet most geomorphological activity will occur where the ice velocity is greatest, close to the equilibrium line (Figure 3.15), and will decrease beneath the ice divide. If the location of the ice divide changes, then the pattern of ice flow within the ice sheet would be reorganised and a new set or population of bedforms will begin to form parallel to the new pattern of ice flow. These would alter or be superimposed on the original set of bedforms (Figure 9.18). Beneath the new ice divide little modification would occur due to low glacier velocity. However, beneath the equilibrium line the old set of bedforms would be quickly eroded and replaced by a new set with an orientation consistent with the new flow pattern. In between these two locations two populations of bedforms coexist in a superimposed or cross-cutting fashion (Figure 9.18). Cross-cutting bedforms therefore hold important information about changing patterns of ice flow.

the low-pressure area extends in front of the sediment ridge. Recent observations suggest that subglacial meltwater flow within the cavity may accentuate the morphology of the flute by eroding sediment along its flanks. Although this model is widely accepted, not all flutes have boulders or bedrock obstacles at their up-ice ends. There are two possible explanations: (i) the boulder was there and has been subsequently removed by ice flow; or (ii) there was never a boulder at the head of the flute and an alternative explanation for the formation of these flutes is required, perhaps in the same way that megaflutes and drumlins form.

In contrast to the formation of flutes there is little consensus about the formation of drumlins, megaflutes and ribbed moraines. The range and diversity of these landforms is so great, particularly in the context of their internal composition, that some researchers have suggested that there may not be a single mechanism responsible for their formation. This concept is known as *equifinality*: different processes form the same morphological products. The acceptance of such an idea should, however, be consequent only upon our failure to find a universal theory. A general model for the formation of drumlins, megaflutes, MSGL and ribbed moraines must be able to explain the following factors.

1. Variables in the theory must be able to account for the different subspecies of subglacial landforms: drumlins, megaflutes, MSGL and ribbed moraines.
2. It must account for the different composition and structure of drumlins, megaflutes, MSGL and ribbed moraines. In particular it must explain the presence of the three main types of drumlin core commonly found: (i) bedrock; (ii) till; and (iii) bedded sands and gravels.
3. It must account for the spatial distribution of bedforms: why do they only occur beneath certain parts of an ice sheet?
4. It must account for the rapid rates of landform creation observed at modern glacier margins.

Most attention has focused on explaining the formation of drumlins. Numerous models, hypotheses and explanations exist but most attention is now focused on the role of subglacial deformation as the most likely explanation. It is important to note, however, that this is not the only idea currently proposed within the literature. In particular the possibility that glacial bedforms are produced by subglacial floods has been a persistent idea over the past decade (Box 9.5), although many consider this to be an 'outrageous' rather than plausible hypothesis.

Observations of subglacial deformation beneath Breiðamerkurökull in Iceland were used in the late 1980s by Geoffrey Boulton and Richard Hindmarsh to develop a flow law with which to describe the deformation of subglacial till. Although this flow law has been widely disputed and amended since its publication, it forms the basis for several models of drumlin formation. The basic concepts of this theory are reviewed below, but it is important to note that it remains simply one of many models (Box 9.6).

When a glacier flows over soft deformable sediment Boulton suggests that three horizons may be identifiable. The surface horizon (A horizon) is rapidly deforming

BOX 9.5: DRUMLINS FORMED BY SUBGLACIAL MELTWATER

Over the past 20 years John Shaw has championed the controversial idea that drumlins are formed by subglacial floods. The idea stems from the similarity in morphology of drumlins with erosional marks at the base of turbdites It rests on the premise that inverted erosional marks at the ice bed are subsequently infilled to form drumlins, as illustrated below. The recognition in the 1980s that many drumlins have a core of sand and gravel gave support to this hypothesis. Significantly, the theory implies that drumlins may not reflect ice-flow direction and that outbursts of subglacial meltwater are critical to landform evolution. Since its inception this theory has, however, proved extremely controversial, attracting fierce debate (Benn and Evans, 2006; Shaw and Munro-Stasiuk, 2006). Its opponents have focused on several key issues: (i) the mechanical implausibility of large areas of an ice sheet or ice margin 'being lifted off' the bed by flood waters and the absence of water sources sufficiently large enough beneath ice sheets to give the required floods; (ii) the

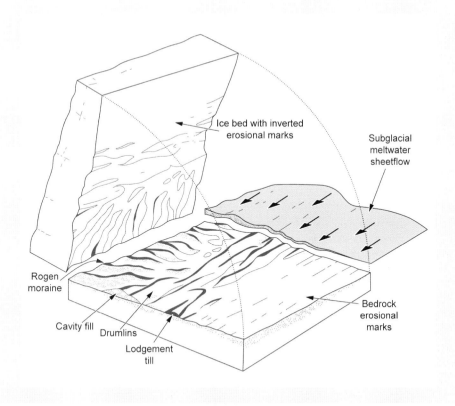

absence of evidence for massive outburst floods other than the drumlins themselves; (iii) the model is not supported by modern analogue observations and drumlins revealed by ice retreat since the Little Ice Age show no evidence of meltwater as a forming process; and (iv) large floods at modern glaciers such as in Iceland have not produced drumlins. Despite these problems some scientists still hold to the idea that subglacial floods provide a possible mechanism of drumlin formation. At its heart the idea challenges the concept of uniformitarinasm – the key to the past is the present – and to accept it one must invoke a different set of glacial processes in the past than operate now.

Source: Shaw, J. (1994) A qualitative view of sub-ice-sheet landscape evolution. *Progress in Physical Geography*, **18**, 159–84. Shaw, J. (2002) The meltwater hypothesis for subglacial bedforms. *Quaternary International*, **90**, 5–22. Benn, D.I. and Evans, D.J.A. (2006) Subglacial megafloods: Outrageous hypothesis or just outrageous? in: *Glacier Science and Environmental Change* (ed. P. Knight), Blackwell, Oxford, pp. 42–6. Shaw, J. and Munro-Stasiuk, M. (2006) Subglacial megafloods: Outrageous hypothesis or just outrageous? Reply, in: *Glacier Science and Environmental Change* (ed. P. Knight), Blackwell, Oxford, pp. 46–50. [Diagram reproduced from: Shaw (2002) *Quaternary International* **90**, figure 1, p. 6. Copyright © 2002, Elsevier Ltd].

BOX 9.6: DRUMLINS AND SUBGLACIAL DEFORMATION

Evidence for the theory of drumlin formation by subglacial deformation is provided by Boyce and Eyles (1991). These authors studied the Peterborough drumlin field in central Canada, which was formed beneath a lobe of ice at the margin of the former Laurentide Ice Sheet during the last glacial cycle. They examined the morphology and internal composition of the drumlins along a line parallel to the direction of glacier flow (a flow line). Along the flow line the drumlins change from elongate to oval forms close to the limit of the ice lobe. The elongated drumlins are composed of till resting on bedrock, whereas the oval forms are eroded from outwash sediments. The drumlins with cores of outwash sediments are mantled with till, which was derived from the deformation and incorporation of underlying till to form a deformation till. A transition therefore occurs along the flow line from depositional to erosional forms close to the ice limit. The down-ice evolution of drumlin form was interpreted by Boyce and Eyles (1991) as a function of the time available for subglacial deformation during the advance of the ice lobe. The duration of deforming bed conditions was greatest up-glacier, where the drumlins are elongated and where all the pre-existing sediment has been eroded and incorporated into the deformation till that forms the drumlins. Towards the limit of the ice lobe, where subglacial deformation will only have occurred for a short period, the drumlins are more oval in shape with cores of undeformed sediment. There was insufficient time here to produced highly streamlined forms or for subglacial deformation to cut down and remould the outwash sediment. These observations not only support the subglacial deformation model of drumlin formation but provide rare insight into the internal composition of drumlins and illustrate how their morphology can vary along a glacier flow line.

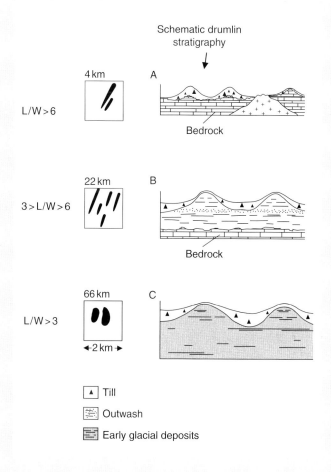

Source: Boyce, J.I. and Eyles, N. (1991) Drumlins carved by deforming till streams below the Laurentide ice sheet. *Geology*, **19**, 787–90. [Modified from: Boyce and Eyles (1991) *Geology*, **19**, figure 2, p. 788.]

and contains material in transport. Beneath this there is a slowly deforming horizon (B_1 horizon). Below this, the sediment is not deforming but is stable (B_2 horizon; Figure 8.13). The boundaries between the three horizons should not be considered to be planar because the thickness of each horizon will vary with the changing properties of the sediment. For example, the presence of the slowly deforming horizon (B_1) is dependant on the *rheology* or stiffness of the sediment. If the sediment is stiff and not easily deformed this horizon may be absent. Sediment rheology is controlled by a range of variables, of which *pore-water pressure* is of particular importance. Pore-water pressure is the pressure of the water within the pores or interstices within the sediment and helps to determine intergranular friction. If the pore-water pressure is high individual grains of sediment are pushed further apart and the friction between one grain and the next is reduced. The lower the level of intergranular friction the more easily the sediment will deform. Fine-grained sediments tends to have a higher pore-water content and pressure than coarse sediments and will therefore deform more easily. Pore-water pressures may also be reduced by increasing the effective normal pressure imposed on a sediment, because this tends to drive off water, provided it can drain away. This increases intergranular friction within the sediment. The shape and size of individual grains of sediment also help determine intergranular friction.

The nature of the boundary between the A and B horizons may either be erosional or depositional depending upon whether the glacier is experiencing extending or compressional flow. As we saw in Section 8.1.1, where a glacier is experiencing extending flow the deforming layer may downcut (erode) into the sediment pile beneath, assimilating new sediment (B horizon) into the deforming layer (A horizon; Figure 8.13). In this case the junction between the A and B horizons is erosional. In areas of compressive flow the deforming layer will grow by the accumulation of till transported laterally in the deforming layer from up-ice areas. In summary, therefore, the nature of the deforming layer is a function of compressive or extending ice flow (Figure 8.6) and of variation in the rheology of the deforming sediment beneath the glacier.

Boulton developed a semi-quantitative flow model for the deformation of the rapidly deforming A horizon on the basis of field observations (see Box 3.4). Using this model he was able to predict how the rapidly deforming A horizon would become moulded around an obstacle to form a drumlin. Figure 9.20A shows the flow lines within a layer of soft deforming sediment. There are two zones of enhanced sediment flow either side of the obstacle and a zone of slower flow in its lee. This pattern of sediment flow produces a sheath of soft sediment around the core as shown in Figure 9.20B. The sediment within this sheath is not stationary, although the shape of the sheath is, because sediment is added at the up-glacier side and removed down-glacier. If the glacier decays and/or the stress field beneath it changes then the deforming A horizon will become stationary around the core to form a drumlin.

The range of drumlin morphologies may be explained by the deformation of this A horizon around either fixed obstacles or mobile obstacles. The obstacles need not, however, necessarily be visible at the surface but simply provide a rigid or stiffer area within the deforming bed. Three types of obstacle were considered in the

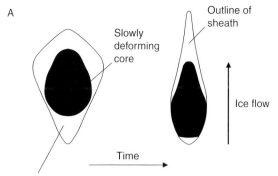

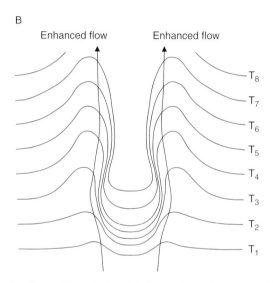

Figure 9.20 Drumlins formed by subglacial deformation. (A) Outline of the shape formed by a sheath of soft deforming sediment around a slowly deforming core. (B) The pattern of flow within a deforming layer passing around a rigid cylinder. The progressive deformation of an originally straight transverse line is followed from T_1 to T_8. Note the reduced rate of deformation in the lee of the cylinder and the enhanced flow along its flanks. [Modified from: Boulton (1987) in: *Drumlin symposium* (eds J. Menzies and J. Rose), Balkema, figure 11, p. 49]

theory: (i) bedrock obstacles (Figure 9.21); (ii) folds within the B_1 horizon (Figure 9.22); and (iii) undeformed areas of sand and gravel (Figure 9.23).

According to Boulton's model, deforming sediment will thicken in front of a bedrock scarp, due to the high effective pressures present, to form drumlinoid noses. These noses form because the high pressures generated by ice flowing against such scarps (see Section 4.6: Figure 4.6) expel water from the sediment and therefore its ability to deform. The reduced rate of deformation (i.e. sediment transport) causes sediment to accumulate. Deforming sediment will also form in

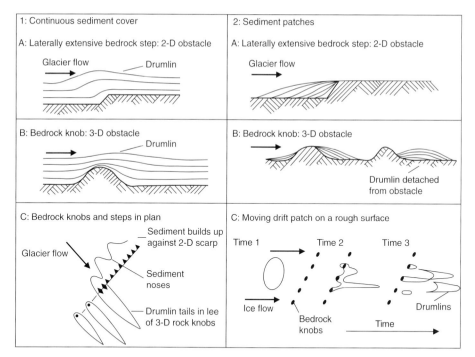

Figure 9.21 Morphology of a deforming layer moving over an irregular bedrock surface. Accumulations of deforming sediment, drumlins, are static where the sediment supply or cover is continuous, but mobile where the sediment cover is patchy and the supply is therefore discontinuous. [Modified from: Boulton (1987) in *Drumlin Symposium* (eds J. Menzies and J. Rose), Balkema, figure 27, p. 72]

the lee of bedrock knobs, due to the decrease in flow of the deforming sediment (Figure 9.21). If the supply of deforming sediment is large then the tail of sediment around the bedrock obstacle will remain, like a stationary or standing wave. In this case there is a constant throughput of sediment within the drumlin; it is added upstream as quickly as it is removed downstream. However, if the supply of deforming sediment is small, for example if it is just a patch of deforming sediment, then the drumlin formed around the bedrock knob will move past the obstacle as sediment is removed from the up-glacier face but not replaced. The pattern of drumlins created by bedrock obstacles depends largely on the availability of sediment and the roughness of the bedrock surface (Figure 9.21).

Folding along the boundary between the B_1 horizon and the A horizon may provide foci for drumlin formation. If the properties of the B_1 horizon vary in the direction of flow, for example if the sediment becomes stiffer down-ice, then its rate of deformation will change (high to low), which may lead to compression and folding. Deformation of the A horizon around such folds may form a drumlin as shown in Figure 9.22. Repeated folding and refolding of the original fold may cause it to be derooted, in the same way that a piece of chewing gum may be stretched and

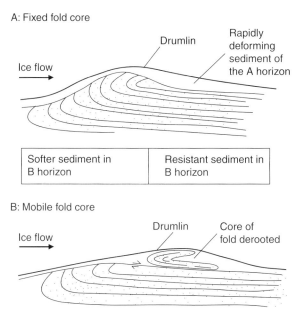

Figure 9.22   Morphology of a deforming layer around a fold generated at the interface between the A and B horizons. (A) Fixed fold core and static drumlin. (B) Derooted, mobile, fold core and therefore mobile drumlin. [Modified from: Boulton (1987) in: *Drumlin Symposium* (eds J. Menzies and J. Rose), Balkema, figure 27, p. 73]

stretched until it finally breaks. Once the fold has been derooted it is able to move in the direction of glacier flow. The drumlin will also be able to migrate.

The final type of drumlin core considered by Boulton was one of undeformed sand and gravel. Figure 9.23 shows a glacier forefield in which coarse gravel is deposited close to the meltwater portal of an ice front. If this ice front was to advance over the outwash sediment it would deform it. The coarse free-draining gravels would be less likely to deform due to low pore-water pressures within them and remain as fixed undeformed cores around which finer grained sediment, less well drained and therefore with higher pore-water pressures, can deform. The core is shaped and eroded by the deformation of the A horizon over its surface and by erosion at the base of this deforming layer. In this way drumlins with undeformed cores of bedded fluvial sands and gravels may be generated (Box 9.5). In areas of strongly extending flow, erosion occurs at the boundary between the A and B horizons as deformation cuts down through the sediment pile. The deforming A horizon may be very thin due to the lateral transport of the deforming sediment assimilated at the A–B horizon. In this case a drumlin may effectively be formed by erosion along the interface between the A and B horizons.

Within this model megaflutes are simply a subtype of drumlin, produced by rapid rates of glacier flow and subglacial deformation, which would tend to produce more elongated forms. Ribbed moraines are considered within this model to be formed by the remoulding of earlier linear bodies of sediment, perhaps formed

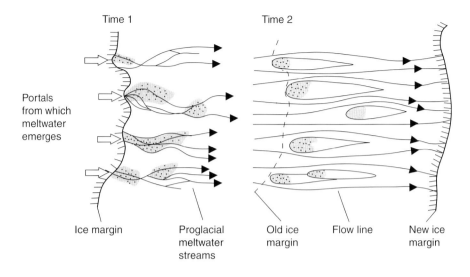

Coarse gravel bars

Figure 9.23   Drumlins initiated around cores of stiff undeforming sediment, in this case coarse gravel bars. Time 1: deposition of coarse gravel close to meltwater portals. Time 2: ice advances and subglacial deformation around the coarse gravel produces streamlined drumlins with cores of undeformed gravel. They are analogous to boudins within a highly deformed rock body. [Modified from: Boulton (1987) in *Drumlin symposium* (eds J. Menzies and J. Rose), Balkema, figure 27, p. 73]

by earlier ice-flow directions, although a range of other explanation have been proposed in recent years (Figure 9.24; Box 9.7).

The strength of Boulton's model lies in the fact that it can explain all the requirements of a general theory, that is the presence of different subspecies of subglacial landforms such as megaflutes, drumlins, MSGL and ribbed moraines. The model also explains the range of different compositions and structure found within these landforms; in particular the presence of drumlin cores composed of: (i) bedrock; (ii) till; and (iii) bedded sands and gravels. Additionally the model also explains the spatial distribution of bedforms: they only occur where subglacial deformation is possible. Finally the model can explain the rapid rates of drumlin formation observed in some studies. The key strength of this model is that it represents an attempt to develop a unified model of drumlin formation by subglacial deformation.

It is not, however, without its opponents. Several researchers have argued that at present there is no direct evidence to suggest that subglacial deformation is a pervasive process beneath ice sheets. To date, direct field observations of subglacial deformation are restricted to fast flowing Antarctic ice streams and glaciers in Iceland and Alaska. Opponents of subglacial deformation point to the absence of widespread evidence of subglacial tectonic structures, such as folds and thrusts within glacial sediments, and to homogeneous till layers and undisturbed sediment sequences. However, as we saw in

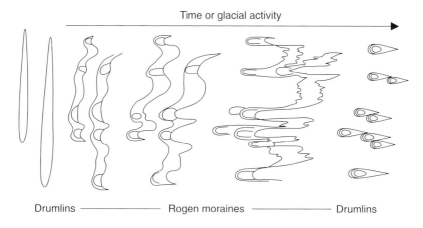

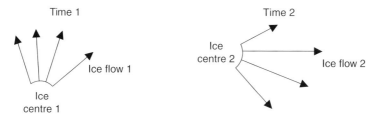

Figure 9.24 Schematic illustration of the way in which transverse drift ridges may be progressively transformed by deformation into rogen or ribbed moraines and drumlins. Original drift ridges, megascale glacial lineations or large drumlins, may reflect an earlier direction of ice flow and are deformed into ribbed moraine by a new ice-flow direction associated with a shift in the ice divide. [Modified from: Boulton (1987) in *Drumlin Symposium* (eds J. Menzies and J. Rose), Balkema, figure 28, p. 75]

BOX 9.7: FORMATION OF ROGEN (RIBBED) MORAINES

Rogen moraines, also referred to as ribbed moraines, occur extensively on the beds of palaeo-ice sheets. They have been explained in a variety of ways, including: (i) as a product of a deforming bed; (ii) due to the fracture of a frozen till sheet at a thermal boundary in an ice sheet; and (iii) due to the deformation of a pre-existing ridge such as a large drumlin (Figure 9.24) or ice-marginal moraine. Of particular interest is the notion that they might be generated due to flow instability within a deforming layer, in the same way that bedforms are created on the bed of a river. This notion works on the idea that small perturbations in the rheology of a deforming layer, essentially an instability, may be amplified to produce large-scale thickness variations – rogen or ribbed moraines – within the deforming layer. These ideas were originally developed to explain drumlins but have been found to provide robust explanations for ribbed moraines. Theoretical models produce land-form wavelengths similar to those observed in the field.

Source: Dunlop, P., Clark, C.D. and Hindmarsh, R.C.A. (2008) Bed ribbing instability explanation: Testing a numerical model of ribbed moraine formation arising from coupled flow of ice and subglacial sediment. *Journal of Geophysical Research*, **113**, F03005, doi:101029/2007JF000954.

Section 8.1.1 intense subglacial deformation is not marked by folds and faults but by homogeneous till units. The absence of widespread tectonic structures does not there-fore provide evidence against subglacial deformation, because they would be present only in areas subject to low levels of deformation.

Others opponents of subglacial deformation have pointed to the juxtaposition of eskers (see Section 9.3) with drumlins. They argue that eskers (formed in subglacial tunnels) could not have formed at a deforming bed where subglacial water flow is believed to occur in shallow canals cut within the sediment. This is only valid provided that: (i) the eskers formed in subglacial cavities; and (ii) they formed at the same time as the drumlins. The eskers may have formed in englacial tunnels and have been subsequently lowered to the glacier bed during deglaciation or alternatively may have formed after the glacier bed had ceased to deform, perhaps during deglaciation. Significantly, work on the distribution of eskers at a continen-tal scale has shown that they are restricted to areas that are unlikely to have experienced subglacial deformation.

BOX 9.8: FORMATION OF BRAIDED ESKERS: SOME EXAMPLES FROM THE SCOTTISH HIGHLANDS

There are a number of very good examples of braided eskers within the Scottish Highlands. In particular several good examples exist within a corridor of eskers, kames and outwash sediments that runs from Lanark in the south-west towards Edinburgh in the northeast. This corridor of glaciofluvial sedi-ment appears to have been deposited in an interlobe environment created by the progressive decoupling, during deglaciation at the end of the last glacial cycle, of two confluent ice sheets within the Central Lowlands of Scotland. To the north was the Highland Ice Sheet and to the south was the ice dome of the Central Uplands. As these ice sheets decoupled progressively from the north-east the area between them became a focus for glaciofluvial sedimentation. Good braided esker systems are to be found at Newbiggin and at Carstairs. The esker system at Newbiggin is described by Bennett *et al.* (2007) and consists of several morphological components, including a single linear esker ridge, inter-rupted and terminated by shallow fans along its length; a series of multiple subparallel ridges and shallow fans; and a complex multiridge structure. The sedimentolgy of the system is consistent with the progressive infilling of a large lake basin by a subglacial conduit, which becomes progressively emergent at the ice surface as a series of ice-walled channels feeding a system of subaqueous fans, as illustrated below. The model demonstrates the complex evolution of glaciofluvial systems along an active ice margin receiving a pronouced meltwater discharge.

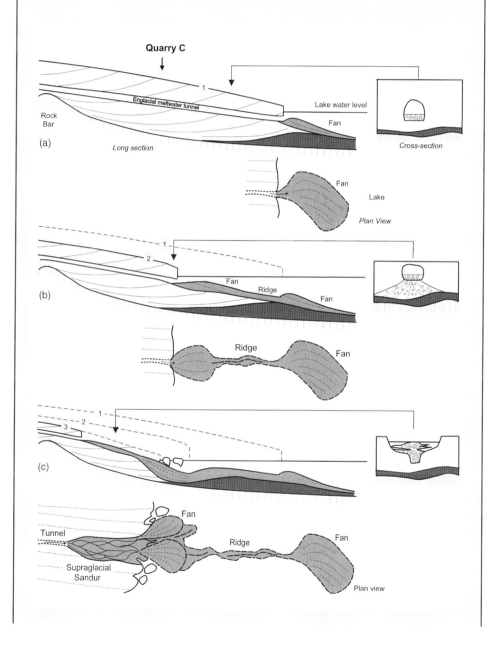

Source: Bennett, M.R., Huddart, D. and Thomas, G.S.P. (2007) in: *The Newbiggin Esker System, Lanarkshire, Southern Scotland: A Model for Composite Tunnel, Subaqueous Fan and Supraglacial Esker Sedimentation* (eds M.J. Hambrey, P. Christoffersen, N.F. Glasser and B. Hubbard) *Glacial Sedimentary Processes & Products, International Association of Sedimentologists Special Publication*, **39**, 177–202. [Modified from: Bennett *et al.* (2007) in: *Glacial Sedimentary Processes & Products* (eds Hambrey *et al.*), *International Association of Sedimentologists Special Publication*, **39**, figure 14, p. 196]

More sophisticated models of bedform formation have also been proposed. These are based on the idea that instabilities within a deforming layer may itself be sufficient for bedform creation (Box 9.7). Bedforms are created by instabilities within many other natural systems (e.g., sand ripples within a river) where there is a mechanism that amplifies small disturbances within a system leading to a regular wavelength of perturbation. The idea is that natural variations in till rheology may lead to formation of similar instabilities within a deforming layer, creating bedforms. Although the computation behind these theoretical models is complex, the basic premise is that bedforms can be generated from instability of flow within a deforming layer. This theory has in recent years found a particular place in explaining the formation of transverse bedforms such as rogen moraines (Box 9.7) and is an area of ongoing research.

9.2.2 Non-Ice Moulded Subglacial Landforms

Subglacial landforms can form in situations where subsequent ice flow is minimal and consequently no ice-moulding occurs. The most important of these landforms are *geometrical ridge networks* or *crevasse-fill ridges* (Figure 9.25). They are low (1–3 m) ridges, symmetrical in cross-section, with a distinct geometrical pattern forming networks when viewed in plan. This geometrical pattern often appears similar to the pattern of crevasses visible at the adjacent ice margin. The ridges are normally composed of basal till. They are believed to form by the squeezing of basal till into subglacial crevasses. If crevasses penetrate to the base of a glacier then basal debris may be squeezed into them and reach an englacial position. Squeezing may also occur into subglacial tunnels. Survival of these ridges is only possible if the ice is stagnant or becomes cold-based immediately after ridge formation. At modern glaciers, geometrical ridge networks are commonly associated with glacier surges. Rapid ice velocities during the surge open up basal crevasses and cause steep thrusts to form, which later become filled with basal sediment as the surge ends. Exceptionally good examples of crevasse-squeeze ridges are to be found on the island of Coraholmen in the Ekman Fjorden fjord, Svalbard. As shown in Figure 9.6 the Sefströmbreen glacier surged forward into the Ekman Fjorden fjord and part of the ice margin became grounded on the island of Coraholmen. As the heavily

Figure 9.25   Crevasse-squeeze ridges formed by a surge of Hagafellsjökull-Eystri (Iceland) in 1999. [Photograph: M.R. Bennett]

crevassed ice-margin settled into the marine mud that it had transported from the floor of Ekman Fjordan onto the limestone island, sediment was squeezed out in front of the ice margin to form a moraine and sediment was squeezed into basal crevasses. Upon deglaciation, a series of crevasse-squeeze ridges was uncovered behind a distinct moraine ridge.

Extensive areas of hummocky moraine around the southern lobes of the former Laurentide Ice Sheet in North America have also been attributed to the upward movement of subglacial sediment (Figure 9.26). Traditionally these were explained by widespread stagnation of ice under a thick supraglacial debris cover melting out

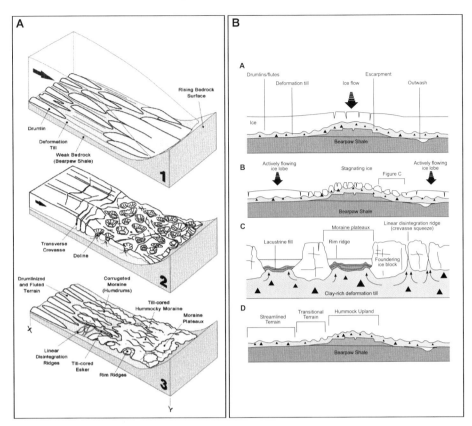

Figure 9.26   Sedimentological model for the formation of hummocky moraine via subglacial deformation. [Reproduced from: Eyles *et al.* (1999) *Sedimentary Geology*, **123**, figures 1, 4 and 5, pp. 164, 168 and 170, Copyright © 1999, Elsevier Ltd]

from debris-rich ice. Amongst this landform assemblage are some distinctive components such as doughnut-like rim ridges, flat-topped moraine plateaux and more linear ridges. Within the Lethbridge Lobe in southern Alberta (Canada) chaotic non-oriented hummocky moraine occurs around the margins of the former ice lobe, which was confined to lower lying terrain. This chaotic terrain passes downslope into weakly oriented hummocks that are transitional to drumlins located in the lobe centres. All these landforms are composed of fine-grained tills with rafts of soft glaciotectonised bedrock derived from the Bearpaw Shales, which underlie the area. A model has recently been put forward which explains this morphological assemblage as a consequence of subglacial 'pressing' into a soft deforming glacier bed during deglaciation (Figure 9.26). The drumlins within the lobe centre record continued active ice flow whereas the hummocky moraine was produced below the outer stagnant margins of the ice lobe by gravitational loading or 'pressing' of dead-ice blocks into wet, plastic till (Figure 9.26). This model has been extended to explain other areas of hummocky moraine below the Laurentide

Ice Sheet and used to support the idea of an extensive deforming subglacial layer. It also provides a challenge to more conventional ideas that hummocky moraine is simply the product of supraglacial sedimentation.

9.3 GLACIOFLUVIAL ICE-MARGINAL LANDFORMS

There are two main sources of glacial meltwater at a glacier snout; water emerging from subglacial meltwater portals or from supraglacial channels fed by surface melt and the emergence of englacial conduits. As glacial meltwater emerges at the ice margin its velocity typically falls due to changes in confining pressure, gradient and increases in bed or channel friction. As a consequence, meltwater streams deposit sediment rapidly to form a variety of *outwash*-related landforms. The morphology and sedimentary composition of these landforms will depend on: (i) the topographic setting of the ice margin and its geometry; (ii) the presence or absence of buried ice; and (iii) the total amount of sediment in transport within the meltwater which is a function of discharge and sediment availability. It is worth noting that some outwash systems are heavily affected by outburst floods, or jökullhaups, generated by subglacial volcanism in volcanically active regions, or by the breaching of ice- and moraine-dammed lakes. These high-magnitude but low-frequency events have a profound impact on the evolution of outwash sequences.

The presence or absence of buried ice and the rate at which it melts is an important control on the morphology of ice-marginal glaciofluvial landforms. As a glacier retreats the drainage pattern will change and on the outwash surface river channels and bars are abandoned. The final morphology of these abandoned outwash surfaces depends upon the amount of buried ice beneath them and whether this ice has melted out before they were abandoned. The conceptual model in Figure 9.27 develops this idea further and shows two glacier margins at which an outwash surface is being deposited. As the two glaciers retreat, the drainage system evolves and this outwash surface is abandoned. At the glacier margin with no buried ice, the outwash morphology reflects the fluvial depositional processes that deposited it, displaying bars, channels and river terraces. This contrasts with the ice margin where buried ice is present. Here, outwash morphology may follow two different evolutionary paths depending upon whether the buried ice has melted out before or after the outwash surface is abandoned. If the ice melts out after the fan is abandoned then the outwash surface will be deformed by subsidence. This may be confined to the occasional *kettle hole* – an enclosed hollow formed by the meltout of buried ice – if the proportion of buried ice is small, but if it is large then the whole surface will be deformed into an area of *kame and kettle topography*. If the meltout of the buried ice is complete prior to the outwash surface being abandoned then there will be little evidence of subsidence on the surface, although it will be evident in the sedimentary structures of the landform (Figure 8.19). As kettle holes form, the meltwater streams will tend to be diverted into them. Deposition will proceed rapidly within these kettle holes because standing water within them causes a reduction in the velocity of stream flow. In this way areas of subsidence are infilled as they form. The two cases illustrated in Figure 9.27

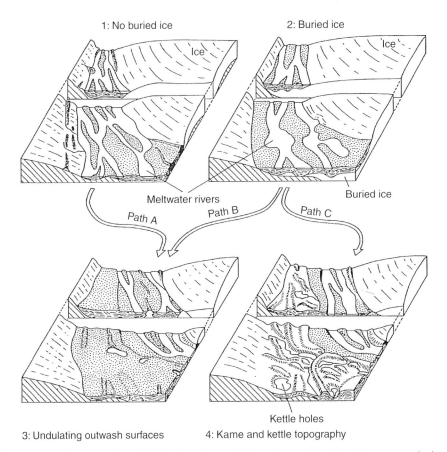

Figure 9.27   Conceptual model showing the evolution of two outwash surfaces, one of which is underlain by buried ice the other is not. The key control on the outwash morphology that results is the timing between the meltout of the buried ice and the abandonment of the outwash surface. If the surface is abandoned before all the ice has melted out then a kame and kettle topography results. Alternatively if the buried ice melts out before the surface is abandoned its presence may not be visible in the outwash surface because meltwater streams tend to infill the kettle holes as they form.

illustrate end-members of a continuum. The character of the resulting landsurface depends on: (i) the rate at which the buried ice melts; (ii) the rate of glacier retreat and the rate at which the outwash surfaces are abandoned; and (iii) the rate of fluvial deposition and its distribution across the outwash surface. Two groups of outwash landforms therefore can be recognised: (i) outwash plains and fans; and (ii) kames and kame terraces.

Outwash fans build up in front of a stationary ice margin (Figures 9.28, 9.29 and 9.30). The apex of each fan is focused on the point at which the meltwater emerges. Fans develop because coarse material is deposited relatively close to the meltwater portal, whereas the finer fraction is transported further. Outwash fans may merge away from the glacier and grade into large braided river sequences forming an

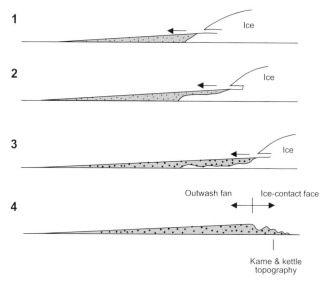

Figure 9.28   The formation and morphology of a simple outwash fan.

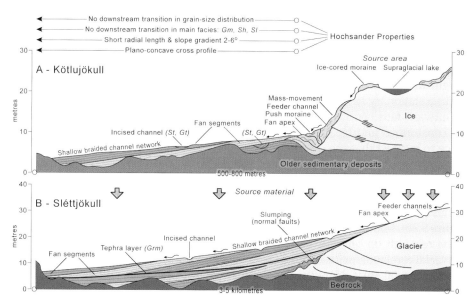

Figure 9.29   Two examples of hochsander or outwash fans in front of outlet glaciers of the Mýrdalsjökull Ice Cap in Iceland. [Modified from: Kjær *et al.* (2004) *Sedimentary Geology*, **172**, figure 13, p. 159]

parallel to the former ice margin and its ice-cored moraines. The sedimentary architecture of kames also reflects their history of subsidence and the removal of lateral ice support. Typical subsidence structures found within kames are illustrated in Figure 9.31.

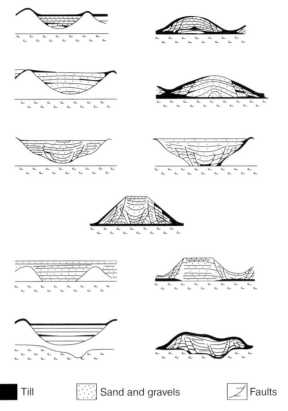

■ Till   ▦ Sand and gravels   ▨ Faults

Figure 9.31 Typical sedimentary structures found within kames. [Modified from: Boulton (1972) *Journal of the Geological Society of London*, **128**, figure 4, p. 370]

So far we have restricted this discussion of glaciofluvial sedimentation to glaciers with simple discharge regimes. However, some glaciers experience more complex discharge regimes, with periodic episodes of catastrophic or jökulhlaup flow (Chapter 4). These high-magnitude, low-frequency discharge events typically introduce large volumes of sediment into a glacier foreland and may have a profound effect on the evolution of glaciofluvial landforms.

9.4 GLACIOFLUVIAL SUBGLACIAL LANDFORMS

The principal landform formed by meltwater flow beneath a glacier is the esker. *Eskers* are the deposits of former subglacial, englacial or supraglacial channels. They are usually slightly sinuous ridges of glaciofluvial sediment that undulate in height along their length. Their orientation is controlled by glacier slope and the pattern of water pressure potential within the glacier; they may therefore show little respect for subglacial topography and need not trend downslope.

Two broad types of esker exist: (i) single-ridge eskers; and (ii) braided (anastomosing) eskers. Braided eskers consist of a network of ridges that merge and bifurcate. They are typically short in length, with braided reaches less than 1 km long, and they are often associated with areas of kame and kettle topography. Single-ridge eskers vary from less than 1 km to hundreds of kilometres in length. Long eskers are typically 400–700 m wide and 40–50 m high, whereas smaller eskers (< 300 m long) are usually 40–50 m wide and only 10–20 m high.

In general eskers may vary in cross-profile along their length and may occasionally have a beaded form. This beaded form may simply consist of wider sections at regular intervals along the length of the esker or alternatively consist of a chain of short lengths of esker ridge between which the ridge is barely visible.

In a few cases eskers have been recorded on top of, and therefore infilling, subglacial N-channels. For example, quarry excavations within the Blakeney Esker in Norfolk, England, reveal a series of small meltwater channels cut into the till surface on which the esker rests. This indicates that the esker formed subglacially and that deposition followed a period of subglacial meltwater erosion. Esker-like bodies of sediment have also been recorded from within units of lodgement till (see Section 8.1.1: Figure 8.12).

Eskers are typically composed of a core of poorly sorted sand and gravels. Above this core, sorted sands and gravels may occur. Sometimes these have arched bedding that dips outwards from the centre of the ridge. Sands and gravels are usually well rounded and have palaeocurrent orientations parallel to the trend of the esker. The esker may be capped by a thin veneer of till. In general, however, the sedimentology of eskers is highly variable and generalisations are difficult. This perhaps reflects the variety of depositional environments in which they form and the high flow regimes or energy levels present.

Single-ridge eskers form when a supraglacial, englacial or subglacial channel or tunnel becomes blocked. Sedimentation in supraglacial channels is easily understood, but sedimentation within englacial or subglacial tunnels is more problematic. By applying theories developed to explain the flow of solids within pipes it has been suggested that sediment of all sizes may be transported as a single mass within englacial or subglacial tunnels (see Section 8.2). The high concentration of debris has a buoyant effect allowing larger particle to be transported more easily. Deposition of these particles is impossible because any constriction of the tunnel caused by deposition would simply increase the water velocity and thereby re-entrain the sediment. However, deposition of all the material in transit takes place if flow within the pipe is suddenly blocked. Obstruction of the tunnel by an ice fall or similar blockage will cause the deposition of large volumes of poorly sorted sand and gravel. Finer grained arched gravel and sands are predicted by pipe flow theory when discharge falls, a situation that would occur at the end of the ablation season. It has been suggested that eskers can form very rapidly due to the high sediment loads within meltwater streams. In this way an esker could form rapidly by the deposition of gravel masses associated with permanent or temporary blockage of a tunnel and by the deposition of arched gravel units during declining discharge at the end of the melt season. Subsequently the esker is uncovered by ice retreat, or lowered to the ground surface by ablation if the tunnel is located in an englacial position. Modification of the esker during lowering is likely, although observations in Iceland suggest that englacial tunnel fills can be lowered quickly with little remobilisation of the esker sediment as the ice retreats. Similarly, supraglacial channel fills can be infilled with sediment and lowered during ablation.

The formation of braided eskers is a more complex problem because it is difficult to envisage the formation of a braided network of subglacial or englacial tunnels. Examples of contemporary eskers from Iceland, however, formed by the infill of englacial and supraglacial tunnels show simple bifurcation. It has also been suggested that subglacial braided channels and associated eskers may form as a result of catastrophic floods. This model suggests that when a single channel cannot accommodate the high discharges of water and sediment during a flood event new channels are produced causing a multichannelled subglacial system to develop, although the exact mechanisms of this process remains unclear. An alternative hypothesis is that braided eskers develop supraglacially either by the development through time of a cross-cutting channel pattern or by the lowering of a supraglacial outwash fan that has deeply incised sediment-filled channels, which are inverted during meltout to form a braided esker.

Observations from within braided eskers in the Southern Uplands of Scotland support this last idea. Here there are two large braided esker systems; one centred at Carstairs and the other at Newbiggin (Box 9.8). The Carstairs system is over 7.5 km long and several hundred metres wide. The major ridges are generally 10–15 m high, but may locally increase in height to over 30 m. The ridges are sharp crested and steep sided. The principal ridges lie within a broad belt of kame and kettle topography, in which low sinuous ridges can occasionally be identified. Internally the main ridges are composed of a core of boulder-rich gravel in which 40–50% of the material is over 250 mm in size and boulders up to 2 m in diameter are also common. The surrounding kames contain much finer grained sands and gravels more typical of that found on braided outwash surfaces. Faulting within these mounds suggests that the sediment has undergone subsidence associated with the meltout of buried ice. It has been suggested that the whole sequence formed as a supraglacial outwash fan subject to catastrophic or high-magnitude low-frequency flood events. During periods of normal flow a system of channels and bars developed on the ice-cored fan surface in which fine gravel and sands were deposited. This was followed by a period of catastrophic flow, which cut a series of deep channels into the buried ice beneath the fan. These channels were then filled with coarse boulder gravel as flow magnitude fell. With the return of normal flow conditions channels were infilled by further low-magnitude deposits. When the outwash surface was abandoned, perhaps as a consequence of the catastrophic flows, meltout of the buried ice inverted the topography. The large boulder-filled flood channels were inverted to form the main esker ridges, while the small channel fills produced the kame and kettle topography around them.

In summary eskers may form in a variety of different settings. These can be summarised as: (i) deposition in subglacial tunnels; (ii) deposition in englacial tunnels and subsequent lowering; (iii) deposition in supraglacial channels and subsequently lowering; and (iv) deposition in ice-walled re-entrants at the ice margin.

9.5 SUMMARY

A wide range of landforms are produced by direct glacial deposition on land. These landforms can be divided into two main groups: those that form by the direct action of glacier ice and those that form by deposition from glacial meltwater. Both groups of landforms can be subdivided into those that form along the ice margin and those that form in a subglacial position. The best known ice-marginal landforms, moraines, form by a combination of processes, which include the pushing, dumping and meltout of glacial debris. The interaction of these processes determines the morphology of the moraine produced, that is a push moraine, dump moraine or ablation moraine. Subglacial landforms are produced beneath a glacier and generally have a streamlined form. Subglacial bedforms include flutes, megaflutes, drumlins, megascale glacial lineations and rogen (ribbed) moraines. They probably form through the processes of subglacial deformation. Non-streamlined landforms, such as crevasse-squeeze ridges, form beneath stagnant or inactive ice.

When meltwater emerges from a glacier, rapid sedimentation can occur along its front and its margin. The morphology of the landforms produced depends upon: (i) whether deposition occurs over the ice margin and thereby incorporates buried ice; (ii) the rate at which this buried ice melts out in relation to when the outwash surface is abandoned due to glacier retreat; and (iii) the rate and distribution of glaciofluvial deposition over the surface. A range of glaciofluvial landforms form at the ice margin, including outwash plains, outwash fans, kame terraces and kames. Glaciofluvial deposition may also occur subglacially to produce eskers. Eskers form either through infill of subglacial and englacial tunnels or by sedimentation in supraglacial channels.

SUGGESTED READING

The volume edited by Evans (2003) on glacial landsystems contains valuable information on a range of different landforms and common landform assemblages. Evans and Twigg (2002) provide an excellent review of the landforms typical of active temperate glaciers. The morphology and formation of glaciotectonic moraines is reviewed by van der Wateren (1995) and by Bennett (2001). The formation of small seasonal push moraines is described by Sharp (1984), Boulton (1986), Humlum (1985) and Krüger (1985, 1993). More complex push moraines are described in papers by Matthews *et al.* (1979), Croot (1987), Hambrey and Huddart (1995), Boulton *et al.* (1996, 1999) and Benediktsson *et al.* (2008). Thrust block moraines, such as that at the Thompson Glacier on Axel Heiberg Island, are described by Evans and England (1991).

The formation of lateral moraines by dumping is covered in the following papers: Humlum (1978), Small (1983), Small *et al.* (1984) and Lyså and Lønne (2001). Cross-valley contrasts in moraine size are discussed in papers by Matthews and Petch (1982), Benn (1989) and Shakesby (1989). Rains and Shaw (1981) describe the formation of moraines by ablation at cold-based glaciers. Ablation moraines are discussed by Goldthwait (1951), Weertman (1961), Souchez (1967), Hooke (1970), and Fitzsimons (1990). The classic literature on hummocky moraines includes Hoppe (1952), and Gravenor and Kupsch (1959). Eyles (1979) provides a modern perspective on hummocky moraine as a supraglacial deposit, and Eyles *et al.* (1999) and Boone and Eyles (2001) discuss its origin in the context of subglacial deformation beneath stagnant ice.

The morphology and formation of flutes is discussed in detail in the paper by Boulton (1976) and in the following papers: Hoppe and Schyatt (1953), Rose (1989), Gordon *et al.* (1992) and Benn (1994). The literature on drumlins is vast, but much of the early literature has been reviewed and compiled by Menzies (1979, 1984). The volume edited by Menzies and Rose (1987) is of particular note and includes the seminal paper by Boulton (1987). The significance of superimposed drumlins and cross-cut patterns is explored by Boulton and Clark (1990a,b), Clark (1993, 1994) and Stokes and Clark (2001). Some pertinent field observations are contained within the papers by Krüger and Thompson (1985), Krüger (1987), Boyce and Eyles (1991), and Eyles and Boyce (1998). The formation of crevasse-squeeze ridges is covered by Sharp (1985).

The characteristics of outwash fans are discussed by Price (1969, 1971), Boulton (1986) and Krüger (1997). Kjær *et al.* (2004) provide an excellent case study of the outwash fans at the margins of the Mýdalsjökull Ice Cap in Iceland, and the paper by Krzyszkowski and Zieliński (2002) is also of particular note. Zielinski and van Loon (2002, 2003) discuss the difference between modern outwash fans and Pleistocene outwash plains in Poland. Boulton (1972) describes the development of kames and their characteristic sedimentary architecture. The impact of jökulhla-ups on outwash surfaces is covered in papers by Desloges and Church (1992), Maizels (1992) and Russell (1994). The kame terraces of Loch Etive, Scotland are discussed by Gray (1975).

The morphology and formation of eskers are discussed in papers by Price (1966, 1969), Shreve (1985), Syverson *et al.* (1994), and Warren and Ashley (1994). The characteristics and formation of braided eskers in the Moray Firth area of Scotland are discussed by Auton (1992). The internal composition of eskers is discussed by Banerjee and McDonald (1975) and the presence of N-channels beneath the Blakeney Esker in Norfolk, England is discussed by Gray (1988).

Aber, J.S. and Ber, A. (2007) *Glaciotectonism. Developments in Quaternary Science*, **6**, Elsevier, Amsterdam.

Auton, C.A. (1992) Scottish landform examples—6: The Flemington eskers. *Scottish Geographical Magazine*, **108**, 190–6.

Banerjee, I. and McDonald, B.C. (1975) Nature of esker sedimentation, in: *Glaciofluvial and Glaciolacustrine Sedimentation* (eds A.V. Jopling and B.C. McDonald), The Society of Economic Paleontologists and Mineralogists, Special Publication 23, Tulsa, pp. 132–542.

Benn, D.I. (1989) Debris transport by Loch Lomond Readvance glaciers in Northern Scotland: Basin form and the within-valley asymmetry of lateral moraines. *Journal of Quaternary Science*, **4**, 243–54.

Benn, D.I. (1994) Fluted moraine formation and till genesis below a temperate valley glacier: Slettmarkbreen, Jotunheimen, southern Norway. *Sedimentology*, **41**, 279–92.

Bennett, M.R. (2001) The morphology, structural evolution and significance of push moraines. *Earth Science Reviews*, **53**, 197–236.

Benediktsson, Í.Ö., Möller, P., Ingólfsson, Ó. *et al.* (2008) Instantaneous end moraine and sediment wedge formation during the 1890 glacier surge of Brúarjökull, Iceland. *Quaternary Science Reviews*, **27**, 209–34.

Boone, S.J. and Eyles, N. (2001) Geotechnical model for Great Plains hummocky moraine formed by till deformation below stagnant ice. *Geomorphology*, **38**, 109–24.

Boulton, G.S. (1972) Modern Arctic glaciers as depositional models for former ice sheets. *Journal of the Geological Society of London*, **128**, 361–93.

Boulton, G.S. (1976) The origin of glacially fluted surfaces: Observations and theories. *Journal of Glaciology*, **17**, 287–309.

Boulton, G.S. (1986) Push-moraines and glacier-contact fans in marine and terrestrial environments. *Sedimentology*, **33**, 677–98.

Boulton, G.S. (1987) A theory of drumlin formation by subglacial deformation, in: *Drumlin Symposium* (eds J. Menzies and J. Rose), Balkema, Rotterdam, pp. 25–80.

Boulton, G.S. and Clark, C.D. (1990a) A highly mobile Laurentide ice sheet revealed by satellite images of glacial lineations. *Nature*, **346**, 813–17.

Boulton, G.S. and Clark, C.D. (1990b) The Laurentide ice sheet through the last glacial cycle: the topology of drift lineations as a key to the dynamic behaviour of former ice sheets. *Transactions of the Royal Society of Edinburgh*, **81**, 327–47.

Boulton, G.S. and Eyles, N. (1979) Sedimentation by valley glaciers: a model and genetic classification, in: *Moraines and Varves* (ed. C. Schlüchter), Balkema, Rotterdam, pp. 11–23.

Boulton, G.S., Van der Meer, J.J.M., Beer, D.J., *et al.* (1999) The sedimentary and structural evolution of a recent push moraine complex: Holmstrømbreen, Spitsbergen. *Quaternary Science Reviews*, **18**, 339–71.

Boulton, G.S., Van der Meer, J.J.M., Hart, J.K., *et al.* (1996) Till and moraine emplacement in a deforming bed surge – an example from a marine environment. *Quaternary Science Reviews*, **15**, 961–87.

Boyce, J.I. and Eyles, N. (1991) Drumlins carved by deforming till streams below the Laurentide ice sheet. *Geology*, **19**, 787–90.

Clark, C.D. (1993) Mega-scale glacial lineations and cross-cutting ice flow landforms. *Earth Surface Processes and Landforms*, **18**, 1–29.

Clark, C.D. (1994) Large-scale ice-moulding: A discussion of genesis and glaciological significance. *Sedimentary Geology*, **91**, 253–68.

Croot, D.G. (1987) Glacio-tectonic structures: A mesoscale model of thin-skinned thrust sheets? *Journal of Structural Geology*, **9**, 797–808.

Desloges, J.R. and Church, M. (1992) Geomorphic implications of glacier outburst flooding: Noeick River valley, British Columbia. *Canadian Journal of Earth Science*, **29**, 551–64.

Dunlop, P. and Clark, C.D. (2006) The morphological characteristics of ribbed moraine. *Quaternary Science Reviews*, **25**, 1668–91.

Dunlop, P., Clark, C.D. and Hindmarsh, R.C.A. (2008) Bed ribbing instability explanation: Testing a numerical model of ribbed moraine formation arising from coupled flow of ice and subglacial sediment. *Journal of Geophysical Research*, **113**, F03005, doi:101029/2007JF000954.

Evans, D.J.A. and England,J. (1991) High Arctic thrust block moraines: Canadian landform examples – 19. *Canadian Geographer*, **35**, 93–7.

Evans, D.J.A. and Twigg, D.R. (2002) The active temperate glacial landsystem: A model based on Breiðamerkurkull and Fjallsjökull, Iceland. *Quaternary Science Reviews*, **21**, 2143–77.

Evans, D.J.A. (ed.) (2003) *Glacial Landsystems*, Arnold, London.

Eyles, N. (1979) Facies of supraglacial sedimentation on Iceland and Alpine Temperate Glaciers. *Canadian Journal of Earth Science*, **16**, 1341–61.

Eyles, N., Boyce, J.I. and Barendregt, R.W. (1999) Hummocky moraine: Sedimentary record of stagnant Laurentide ice sheet lobes resting on soft beds. *Sedimentary Geology*, **123**, 163–74.

Eyles, N. and Boyce, J.I. (1998) Kinematics indicators in fault gouge: Tectonic analog for soft-bedded ice sheets. *Sedimentary Geology*, **116**, 1–12.

Fitzsimons, S.J. (1990) Ice-marginal depositional processes in a polar maritime environment, Vestfold Hills, Antarctica. *Journal of Glaciology*, **36**, 279–86.

Goldthwait, R.P. (1951) Development of end moraines in east-central Baffin Island. *Journal of Geology*, **59**, 567–77.

Gordon, J.E., Whalley, W.B., Gellatly, A.F. and Vere, D.M. (1992) The formation of glacial flutes: Assessment of models with evidence from Lyngsdalen, North Norway. *Quaternary Science Review*, **11**, 709–31.

Gravenor, C.P. and Kupsch, W.O. (1959) Ice disintegration features in western Canada. *Journal of Geology*, **67**, 48–64.

Gray, J.M. (1975) The Loch Lomond Readvance and contemporaneous sea-levels in Loch Etive and neighbouring areas of western Scotland. *Proceedings of the Geologists' Association*, **86**, 227–38.

Gray, J.M. (1988) Glaciofluvial channels below the Blakeney Esker, Norfolk. *Quaternary Newsletter*, **55**, 8–12.

Hambrey, M.J. and Huddart, D. (1995) Englacial and proglacial glaciotectonic processes at the snout of a thermally complex glacier in Svalbard. *Journal of Quaternary Science*, **10**, 313–26.

Harris, C. Brahbham, P.J., Williams, G.D. and Eaton, G. (1997) The nature and significance of glaciotectonic structures at Dinas Dinlle, northwest Wales UK. *Quaternary Science Reviews*, **16**, 109–27.

Hattestrand, C. and Kleman, J. (1999) Ribbed moraine formation. *Quaternary Science Reviews*, **18**, 43–61.

Hewitt, K. (1967) Ice-front deposition and the seasonal effect: A Himalayan example. *Transaction of the Institutes of British Geographers*, **42**, 93–106.

Hooke, R. Le, B. (1970) Morphology of the ice-sheet margin near Thule, Greenland. *Journal of Glaciology*, **2**, 140–4.

Hoppe, G. and Schyatt, V. (1953) Some observations on fluted moraine surfaces. *Geografiska Annaler*, **35**, 105–15.

Hoppe, G. (1952) Hummocky moraine regions, with special reference to the interior of Norrbotten. *Geografiska Annaler*, **34**, 1–72.

Humlum, O. (1978) Genesis of layered lateral moraines: Implications for palaeoclimatology and lichenometry. *Geografisk Tidsskrift*, **77**, 65–72.

Humlum, O. (1985) Genesis of an imbricate push moraine, Höfdabrekkujökull, Iceland. *Journal of Geology*, **93**, 185–95.

Kjær, K.H. and Kruger, J. (2001) The final phase of dead-ice moraine development: processes and sediment architecture, Kötlujökull, Iceland. *Sedimentology*, **48**, 935–52.

Kjær, K.H., Sultan, L. Kruger, J. and Schomacker, A. (2004) Architecture and sedimentation of outwash fans in front of the Mýrdalsjökull ice cap, Iceland. *Sedimentary Geology*, **172**, 139–63.

Krüger, J. and Thomsen, H.H. (1984) Morphology, stratigraphy, and genesis of small drumlins in front of the glacier Mýrdalsjökull, south Iceland. *Journal of Glaciology*, **30**, 94–105.

Krüger, J. (1985). Formation of a push moraine at the margin of Höfdabrekkujökull, south Iceland. *Geografiska Annaler*, **67a**, 199–212.

Krüger, J. (1987) Relationship of drumlin shape and distribution to drumlin stratigraphy and glacial history, Mýrdalsjökull, Iceland, in: *Drumlin Symposium* (eds J. Menzies and J. Rose), Balkema, Rotterdam, pp. 257–66.

Krüger, J. (1993) Moraine-ridge formation along a stationary ice front in Iceland. *Boreas*, **22**, 101–9.

Krüger, J. (1997) Development of minor outwash fans at Kötlujökull, Iceland. *Quaternary Science Reviews*, **16**, 649–59.

Krzyszkowski, D. and Zieliński, T. (2002) The Pleistocene end moraine fans: Controls on their sedimentation and location. *Sedimentary Geology*, **149**, 73–92.

Lyså, A. and Lønne, I. (2001) Moraine development at a small High-Arctic valley glacier: Rieperbreen, Svalbard. *Journal of Quaternary Science*, **16**, 519–29.

Maizels, J. (1992) Boulder ring structures produced during Jökulhlaup flow. *Geografiska Annaler*, **74**A, 21–33.

Matthews, J.A. and Petch, J.R. (1982) Within-valley asymmetry and related problems of neoglacial lateral moraine development at certain Jotunheimen glaciers, southern Norway. *Boreas*, **11**, 225–4.

Matthews, J.A., Cornish, R. and Shakesby, R.A. (1979) Saw-tooth moraines in front of Bødalsbreen, southern Norway. *Journal of Glaciology*, **22**, 535–46.

Matthews, J.A., McCarroll, D. and Shakesby, R.A. (1995) Contemporary terminal moraine ridge formation at a temperate glacier: Styffedalsbreen, Jotunheimen, southern Norway. *Boreas*, **24**, 129–39.

Menzies, J. (1979) A review of the literature on the formation and location of drumlins. *Earth Science Review*, **14**, 315–59.

Menzies, J. and Rose, J. (1987) *Drumlin Symposium*, Balkema, Rotterdam.

Price, R.J. (1966) Eskers near the casement glacier, Glacier Bay, Alaska. *Geografiska Annaler*, **48**Λ, 111–25.

Price, R.J. (1969). Moraines, sandur, kames and eskers near Breiðamerkurjökull, Iceland. *Transactions of the Institute of British Geographers*, **46**, 17–43.

Price, R.J. (1970) Moraines at Fjallsjökull. *Journal of Arctic and Alpine Research*, **2**, 27–42.

Price, R.J. (1971) The development and destruction of a Sandur, Breiðamerkurjökull, Iceland. *Journal of Arctic and Alpine Research*, **3**, 225–37.

Price, R.J. (1973) *Glacial and Fluvioglacial Landforms*, Oliver & Boyd, Edinburgh.

Rains, R.B. and Shaw, J. (1981) Some mechanisms of controlled moraine development, Antarctica. *Journal of Glaciology*, **27**, 113–28.

Rose, J. (1989) Glacier stress patterns and sediment transfer associated with the formation of superimposed flutes. *Sedimentary Geology*, **62**, 151–76.

Russell, A.J. (1994) Subglacial Jökulhlaup deposition, Jotunheimen, Norway. *Sedimentary Geology*, **91**, 131–44.

Shakesby, R.A. (1989) Variability in neoglacial moraine morphology and composition, Storbreen Jotunheim Norway: Within-moraine patterns and the implications. *Geografiska Annaler*, **71**A, 17–29.

Sharp, M. (1984) Annual moraine ridges at Skálafellsjökull, South-east Iceland. *Journal of Glaciology*, **30**, 82–93.

Sharp, M. (1985) Crevasse-fill ridges – a landform type characteristic of surging glaciers? *Geografiska Annaler*, **67**A, 213–20.

Shreve, R.L. (1985) Esker characteristics in terms of glacier physics, Katahdin esker system, Maine. *Geological Society of America Bulletin*, **96**, 639–46.

Small, R.J. (1983) Lateral moraines of glacier de Tsidjiore Nouve: Form, development, and implications. *Journal of Glaciology*, **29**, 250–9.

Small, R.J., Beecroft, I.R. and Stirling, D.M. (1984) Rates of deposition on lateral moraine embankments, Glacier de Tsidjiore Nouve, Valais, Switzerland. *Journal of Glaciology*, **30**, 275–81.

Souchez, R.A. (1967) The formation of shear moraines: An example from South Victoria land, Antarctica. *Journal of Glaciology*, **6**, 837–43.

Stokes, C.R. Clark, C.D. (2001) Palaeo-ice streams. *Quaternary Science Reviews*, **20**, 1437–57.

Syverson, K.M., Gaffield, S.J. and Mickelson, D.M. (1994). Comparison of esker morphology and sedimentology with former ice-surface topography, Burroughs Glacier, Alaska. *Geological Society of America Bulletin*, **106**, 1130–42.

Thomas, G.S.P., Connaughton, M. and Dackombe, R.V. (1985) Facies variation in a late Pleistocene supraglacial outwash sandur from the Isle of Man. *Geological Journal*, **20**, 193–213.

Van der Wateren, D.F.M. (1995) Processes of glaciotectonism, in: *Modern Glacial Environments: Processes, Dynamics and Sediments. Glacial Environments* (ed. J. Menzies), Butterworth-Heineman, London, pp. 309–51.

Warren, W.P. and Ashley, G.M. (1994) Origins of the ice-contact stratified ridges (eskers) of Ireland. *Journal of Sedimentary Research*, **A64**, 433–49.

Weertman, J. (1961) Mechanism for the formation of inner moraines found near the edge of cold ice caps and ice sheets. *Journal of Glaciology*, **3**, 965–78.

Worsley, P. (1974) Recent 'annual' moraine ridges at Austre Okstindbreen, Okstidan, north Norway. *Journal of Glaciology*, **13**, 265–77.

Zielinski, T. and van Loon, A.J. (2002) Present-day sandurs are not representative of the geolgocial record. *Sedimentary Geology*, **152**, 1–5.

Zielinski, T. and van Loon, A.J. (2003) Pleistocene sandur deposits represent braidplains, not alluvial fans. *Boreas*, **32**, 590–611.

10: Glacial Sedimentation in Water

The sediments produced by glaciers terminating in water are different to those deposited on land. If the glacier is *grounded* (i.e. in contact with its bed), as opposed to floating, subglacial sedimentation can occur by a combination of meltout, lodgement and deformation, as in terrestrial environments. At the ice margin, the glacier may become detached from the bed and float; a process that gives rise to an ice shelf if the rate of ice discharge exceeds the rate of ice calving. The processes of deposition at floating ice margins are very different and the glacier simply acts as a source of debris, with the depositional processes being controlled not by the direct agency of glacier ice but through the processes of sedimentation within the water body. In the strict sense of the definition used in Chapter 8 tills cannot be produced in this environment because disaggregation of the debris occurs within the water column, although diamicts – deposits that may look like till – can form. Glaciers can terminate in either lakes (lacustrine) or in marine waters and although the sedimentary products are very similar in both environments, there are a number of subtle process-related differences.

10.1 SEDIMENTATION IN LACUSTRINE ENVIRONMENTS

Lacustrine ice margins develop in a variety of different situations. Glaciers may dam lakes, lakes may develop in front of a glacier due to the melting of stagnant ice beneath the proglacial surface, lakes may be dammed between moraines and an ice front, or a glacier may simply drain into a rock-confined basin. During the last glacial cycle both the Laurentide and Fennoscandinavian ice sheets terminated in large proglacial lakes formed in part by the glacio-isostatic depression of the crust in front of the ice margin. The weight of the ice sheets caused depression of the crust, which was compensated for by broad topographic bulges (*forebulges*) several tens of kilometres beyond the ice margin. Proglacial lakes formed between these

Glacial Geology: Ice Sheets and Landforms   Second Edition   Matthew R. Bennett and Neil F. Glasser
© 2009 John Wiley & Sons, Ltd

bulges and the ice margin. For example large lakes formed in front of the Laurentide Ice Sheet in the area of the Great Lakes today and, as a consequence, glaciolacustrine sediments occur extensively in this area.

The pattern of sedimentation within glacial lakes is controlled by the density stratification within the water body. Water density is controlled primarily by temperature and secondarily by salinity and the suspended sediment content. In the summer most lakes develop a strong water stratification in which a surface layer of warm, and therefore less dense, water (*epilimnion*) sits above a lower body of cold denser water (*hypolimnion*; Figure 10.1). In the autumn this surface layer is cooled rapidly, becomes denser and therefore sinks to the bottom of the lake causing the water body as a whole to mix. This stratification controls the resulting processes of sedimentation, which may occur through any or all of the following eight processes.

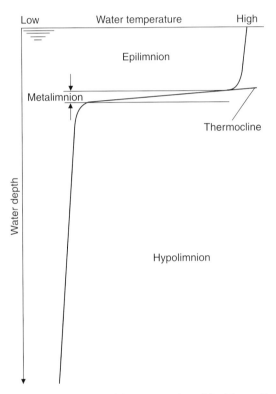

Figure 10.1   The thermal stratification of lake waters. [Modified from: Drewry (1986) *Glacial Geologic Processes*, Arnold, figure 11.3, p. 169]

1. **Deposition from meltwater flows.** The manner in which meltwater enters into a lake depends upon the density of the meltwater relative to that of the lake. If there is a significant difference in density, the sediment-laden meltwater will maintain its integrity as a plume and enter in one of three ways: (i) as an *underflow*, where the sediment plume is denser than the lake water and therefore sinks to the base

of the lake; (ii) as an *interflow*, where the sediment plume is of similar density to the surface water (epilimnion) but less dense than the basal water (hypolimnion) and the plume enters at intermediate depth; and (iii) as an *overflow*, where the sediment is less dense than the surface lakewater and therefore rises to the surface. Meltwater introduced as an overflow or interflow loses immediate traction with the bed and rapid deposition of bedload will occur. These conditions are ideal for the formation of a *delta* – a body of sediment deposited in a water body at a river or stream mouth. As sedimentation occurs a delta will grow away (prograde) from the shore giving a distinctive internal architecture of *topsets*, *foresets* and *bottomsets* (Figure 10.2). The foresets are formed as sediment avalanches down the front face of the delta, while the bottom sets form as a result of fine-grained sediment flows initiated by this process of avalanching and by suspension settling from the water column.

In lakes where there is a pronounced underflow, sediment-laden meltwater descends rapidly into the lake basin as a series of sediment-rich currents. These currents may inhibit the development of large deltas close to the lake margin as they carry much of the sediment into deep water. Persistent underflow eventually develops a subaqueous or lake floor fan. These fans consist of a system of channels, often with lateral levées, which cut into the delta front and feed a series of broad lobes that reach out towards the centre of the lake basin.

2. **Direct deposition from the glacier front.** Material may simply be dumped into the water body from the glacier front. The degree of disaggregation that occurs will depend on the water depth and upon the current activity within the lake. This will result in irregular shaped deposits of diamict.

3. **'Rain-out' from icebergs.** Calving icebergs may float debris out into the body of a lake where it may be released as the iceberg melts (Figure 10.3). The deposits produced are variable, depending upon the density of the debris concentration within the icebergs and the number of icebergs present. Low concentrations of debris and icebergs may result in the deposition of occasional *dropstones*, outsized clasts set in fine-grained sediments (Figure 10.4), whereas high concentrations may result in thick laterally extensive deposits of diamict. Irregularly shaped packages of debris may also be associated with dropstones and are formed by the dumping of sediment from the icebergs as they capsize or as pockets of debris meltout.

4. **Settling from suspension.** Sediment within the main body of a lake will gradually settle to form thin layers of mud and clay that drape other sediments. This type of sedimentation dominates over much of the lake floor away from its margins.

5. **Resedimentation by gravity flows.** Sediment within a lake may become unstable on steep slopes, particularly where rapid deposition occurs. Slumping or flow of this sediment may give rise to a range of diamicts, the properties of which depend on the fluidity of the flow. The greater the flow mobility, the greater the sorting and fabric development within the diamict.

6. **Current reworking.** Currents within the lake may rework and sort sediment that has already been deposited

Figure 10.4   Dropstones in glaciolacustrine deposits in Patagonia. [Photograph: P. Doyle]

whereas fine-grained deposits dominate in the lake centre. In fact we can identify two broad facies within a lake: (i) the basin-margin facies; and (ii) the lake-floor facies.

The basin-margin environment and the associated sedimentary facies are dominated by the inflow of water into the lake. Where water enters as an underflow a higher proportion of coarse sediment is carried further into the lake basin, inhibiting the size of the delta produced and increasing the basinward transport of sediment. This prodelta area may be associated with lobes and fans formed by deposition from subaqueous sediment gravity currents such as turbidity currents. In contrast, where meltwater enters as an overflow or inflow, much of the sediment load is deposited quickly on the delta and the degree of basinward transport is more limited. In practice this may vary seasonally as the temperature and density of the inflow varies along with the degree of water stratification in the lake.

The nature of the processes of sedimentation away from the lake margin depends upon the amount of ice-rafted debris being introduced into the system – in effect the degree to which the lake is connected to the ice margin. The greater the length of the ice margin that is in contact with the lake, the greater the potential for iceberg calving and basinward transfer of ice-rafted debris. We can identify a broad continuum from those lakes in which the input of ice-rafted debris is small, and therefore sedimentation is dominated by settling from suspension and the inflow of meltwater, to those lakes that are dominated by ice-rafted sedimentation. If the ice-rafted component is small then rhythmic fine-grained sediments produced by settling of material out of suspension tend dominate, giving rise to fine laminations of silt and clay (Figure 10.5). This process is interrupted by the deposition of coarser laminations from material introduced by sediment gravity flows (e.g., turbidity

Figure 10.5 A range of glaciolacustrine sediment types. (A) Finely laminated silt and sand units forming rhythmic beds. (B) Dropstone within laminated silts and sands. (C) Waterlain diamict, showing different degrees of current winnowing. (D) Waterlain diamict. [Photographs: M.R. Bennett]

currents), generated either by underflow or by slumping on the delta front or lake margin. This forms rhythmic sediment in which there are coarse- and fine-grained laminations giving a distinctive couplet (Figure 10.5B). When or where underflow is not active, for example during winter months when the inflow or discharge of water into the lake is low, settling from suspension will dominate. Laminations produced in this way typically grade in size from silt and clay particles at the base to fine clay at the top. They are usually terminated by a sharp contact formed by the influx of a new underflow of coarse material. The relative thickness and importance of these two components changes with distance away from the sediment source or point of water inflow. Close to the delta the coarse sand/silt layer, formed by underflow, will be thick. In the basin centre the coarse layers will be much thinner and may be absent. In such cases a continuous deposit of homogeneous fine clay may result. In specific circumstances laminated or rhythmic sediment may have an annual pattern. Underflow and therefore coarse lamination dominate in the summer when the influx of meltwater to a lake is high, whereas suspension settling and fine lamination tend to dominate in the winter when the influx of turbid flows is small or absent. Where an annual cycle can be identified within the rhythmic sediments they are known as *varves*. It is important to emphasise, however, that not all rhythmic lake sediments necessarily contain an annual signature. These rhythmic deposits are easily disturbed or deformed during or after deposition and may show a variety of deformation structures.

In contrast, if the ice-rafted debris component is large, sedimentation is dominated by coarse debris melting from icebergs (Figure 10.5C,D). In this case the deposition of diamicts and resedimentation by gravity flows will dominate. The type of diamict produced in these environments will depend on the interplay between three processes: (i) the rate of debris rainout; (ii) the degree of downslope, basinward remobilisation via sediment gravity flows; and (iii) the degree of current winnowing and reworking on the lake floor. In some of the large glaciolacustrine lakes along the margins of the former mid-latitude ice sheets extensive deposits of diamicts have been deposited in this way (Box 10.1).

BOX 10.1: GLACIAL TILL OR ICE-RAFTED DIAMICTS?

Along the shore of Lake Ontario to the east of Toronto, in Canada, a sequence of glacial sediments is exposed at Scarborough Bluffs. This sequence is dominated by large units of fine-grained diamict separated by bedded and laminated sand and mud units. Traditionally, this sequence has been interpreted as the product of multiple ice advances by a grounded ice sheet. Between each advance lakes formed to give the sandier sediments. Eyles and Eyles (1983) challenged this interpretation on the basis of detailed sedimentological logging. They made the following observations and inferences.

- **Observation:** there were no tectonic structures in the sand and mud units.
  Inference: this is inconsistent with the passage of a grounded ice sheet over the sediment.

- **Observation:** thin stringers of silt and sand are found within the diamicts.
  Inference: water currents were active during the deposition of the diamicts.

- **Observation:** diamict units show evidence of having flowed into and accumulated preferentially within topographic lows.
  Inference: gravity was important in the sedimentation of the diamict units.

- **Observation:** transitional and interbedded contacts occurred between the sand and the diamict units.
  Inference: there was a gradual transition from the deposition of the sandier to the diamict units. This is inconsistent with an ice advance, which would tend to give erosional, non-conformable contacts.

- **Observation:** load structures are present at the contact of the diamict and sandier units.
  Inference: the diamict was deposited onto saturated sands and muds.

- **Observation:** the sandier units contain structures typical of deltas.
  Inference: the sand units were deposited in deltas.

On the basis of these observations and inferences Eyles and Eyles (1983) rejected the traditional model of multiple ice advances. The sediment sequence at Scarborough Bluffs was reinterpreted as being glaciolacustrine, formed by the repeated progradation of deltas over glaciolacustrine diamicts, deposited by both the 'rain-out' of ice-rafted debris and by sediment gravity flows. This interpretation has significant implications for the glacial stratigraphy of the area, in particular for the number of ice advances recognised within the region.

This work represented a significant step forward in glacial geology and opened the way for the reinterpretation of many diamict sequences traditionally viewed simply as the product of grounded glaciers. It also illustrates the process of sediment interpretation, which involves three steps: (i) careful observation/recording; (ii) making inferences from observations; and (iii) examining these inferences to assess the possible explanations or models.

Source: Eyles, C.H. and Eyles, N. (1983) Sedimentation in a large lake: A reinterpretation of the Late Pleistocene stratigraphy at Scarborough Bluffs, Ontario, Canada. *Geology*, **11**, 146–52.

The facies patterns or facies architecture that occurs is largely controlled by the geometry of the lake basin. Important factors include the number of meltwater streams which feed a glacial lake, the presence and geometry of any ice margin present, and the presence of buried ice. The long-term stability of the lake and its evolution are also very important. For example, a supraglacial lake formed on dead ice is highly unstable and constantly prone to change as the ice beneath it melts. As a consequence of the range of possible variables it is often hard to predict the facies patterns which might occur within glaciolacustrine environments (Box 10.2).

BOX 10.2: GLACIAL FACIES ARCHITECTURE ASSOCIATED WITH ICE-DAMMED LAKES IN BRITISH COLUMBIA

As an ice sheet decays and the underlying topography emerges, water frequently accumulates in valleys dammed by ice giving a rapidly changing landscape. This was particularly true of the Cordilleran Ice Sheet in British Columbia as it decayed between 13 000 and 10 000 years BP. Johnsen and Brennand (2006) explored the range of glaciolacustrine facies that accumulated in one such lake within the Thompson Valley, and provide an excellent example of the assemblage of glacial facies that typically characterise such environments. On the basis of detailed description of the exposed sediments within the Thompson Valley they recognise a series of discrete environments:

1. **Delta sediments.** Three large deltas were identified within the valley associated with the inflow of glacial meltwater from retreating ice margins. The topsets of these deltas were typically composed of coarse, imbricate cobble gravel deposited in braided channels on the delta surface. The foresets were found to be composed of both sand and gravel. They record the progradation of different delta lobes into the lake. The sedimentology of these foresets suggests that they were deposited by hyperpycnal flows – cold dense water – flowing over the delta front and out over the lake floor. This clearly indicates that the meltwater inflow was colder and more sediment-rich than the lake water. The foresets grade out into finer-grained bottom sets.

2. **Subaqueous fans.** Nine subaqeuous fans were mapped within the lake, each located where a tributary joins the valley. These bodies of sediment form gently inclined structures which steepen to an apex close to the tributary mouth. They are composed of a range of coarse sands and fine gravels. They occur at the base of the lacustrine fill and are overlain by finer-grained sediments recording a gradual decrease in discharge into the lake as the ice margins feeding the tributaries retreated up-valley. The fans were deposited by a range of subaqueous sediment gravity flows associated with hyperycnal water inflow.

3. **Lake-bottom sediments.** This facies consists of near-horizontal layers of sand and silt showing varying degrees of current traction within a background of suspension settling. Very fine-grained bottom sediments are absent, reflecting the energetic inflow of water to the lake. Interestingly, these bottom sediments are occasionally interbedded with disturbed sediment layers formed by the downslope slumping of sediment from the lake sides and by angular debris derived from valley-side rock falls.

All the facies within the Thompson Valley show evidence of loading and dewatering caused by rapid sedimentation in an unstable and rapidly changing environment. In contrast to other glacial lakes, however, the proportion of ice-rafted diamict and dropstones is small. This reflects the limited extent of the active ice margin that was in contact with the lake. The paper provides a good example of the range of glaciolacustrine facies that exist and the transitory nature of such environments during deglaciation.

Source: Johnsen, T.F. and Brennand, T.A. (2006) The environment in and around ice-dammed lakes in the moderately high relief setting of the southern Canadian Cordillera. *Boreas*, **35**, 106–25.

10.2 SEDIMENTATION IN MARINE ENVIRONMENTS

Glaciers may terminate in the sea either within the confines of a drowned glacial valley, a *fjord*, or in the open sea/ocean. During the Cenozoic Ice Age many of the mid-latitude ice sheets extended towards the edge of their continental shelves, as did the ice sheets of Antarctica and Greenland. During the 1990s there was an increased emphasis on the importance of glaciomarine sedimentation and several glacial sequences traditionally interpreted as being deposited by grounded ice were reinterpreted. Glaciomarine sediments also form common components of the Pre-Cenozoic glacial record because they are preferentially deposited and preserved within marine basins and consequently have greater preservation potential. An understanding of glaciomarine sedimentary processes and products is therefore vital to interpreting the Earth's glacial record (Box 10.3).

BOX 10.3: ONSET OF NORTHERN HEMISPHERE GLACIATION

Marine sediments provide a better record of Earth's glacial history than terrestrial sediments because their rate of sedimentation is more continuous and they are subjected to fewer episodes of erosion. A glacial signature can be found in these sediments via the presence of ice-rafted material such as dropstones or pods of coarse-grained debris within very fine-grained marine silts. This type of evidence has been used recently to challenge the accepted date for the onset of northern hemisphere glaciation. The widely accepted age for the start of glaciation in the northern hemisphere is between 2 and 15 Ma. However, recent marine cores extracted from the Greenland Sea suggest that that glaciation may have started as early as 30–44 Ma in parts of Greenland, because these cores contain evidence of ice-rafted material, which can only have been derived from the calving of icebergs. For this to have happened, glacier ice must have been close to sea level at this time. Glaciation in Antarctica started at a similar time and the new evidence supports the idea that bipolar glaciation occurred in a synchronous manner, perhaps as a result of a global decrease in the greenhouse gas carbon dioxide, which was reduced at this time by increased continental weathering.

Source: Tripati, A.K., Eagle, R.A., Morton, A., *et al.* (2008) Evidence for glaciation in the northern hemisphere back to 44 Ma from ice-rafted debris in the Greenland Sea. *Earth Planetary Science Letters*, **265**, 112–22.

Glaciomarine sedimentation involves a similar range of processes to sedimentation in glacial lakes. However, there are a number of key differences due to the salinity of the water column and the influence of tides. Key processes of glaciomarine sedimentation include the following.

ice-rafted debris, which is followed by the deposition of a lodgement or deformation till as the area is covered by the advancing ice. Deglaciation is first marked by the development of a grounding-line fan and subsequently by a gradual decrease in the amount of ice-rafted debris as the glacier retreats and the glacial influence falls. In practice the pattern of sedimentary facies associated with the advance and retreat of a glacier across a continental shelf may be much more complex.

Within fjords the facies architecture depends on the dynamics and behaviour of the glacier and therefore may be even more complex than that found on the continental shelf. Work on fjord glaciers in Alaska has identified five types of glacier behaviour, each associated with its own pattern of glaciomarine facies.

1. **Association I: rapid deep water retreat.** Here the glacier is retreating rapidly in deep water by iceberg calving. In the ice-proximal zone the sediment facies consists of reworked subglacial till and glaciofluvial sands and gravels exposed as the glacier retreats. These are associated with supraglacial debris dumped from the ice margin. The proximal and distal glaciomarine zones contain large amounts of ice-rafted debris.

2. **Association II: stabilised or slowly retreating ice margin.** Here the ice recession has been retarded by a fjord constriction or pinning point (see Figure 3.2). Calving continues, however, and thick deposits of ice-rafted sediment accumulate. The ice margin is marked by the deposition of coarse-grained sediment either in a grounding-line fan, moraine bank or ice-contact delta (see Section 11.2).

3. **Association III: slow retreat in shallow water.** Here the glacier is either retreating or slowly advancing in shallow water. Calving is severely reduced and so there is relatively little ice-rafted sediment. An ice-contact delta may develop at the ice margin.

4. **Association IV: proximal terrestrial glacier.** By this point the glacier is terrestrial, producing a large outwash delta that progrades into the fjord. The resulting facies are coarse-grained sediments on the delta top, while the delta fronts comprises sand and gravel, which alternate with marine muds deposited from suspension. There is little ice-rafted debris beause only small icebergs are introduced into the fjord via meltwater streams. Sand and silt rhythmites may be deposited on the fjord floor due to the interplay of sedimentation from suspension and turbidity currents generated from the delta slope.

5. **Association V: distant terrestrial glacier.** By now the glacier is distant and so the facies comprise tidal-flat muds and braided stream gravels

These facies associations can be combined in a variety of different ways depending on: (i) the morphology of the fjord; and (ii) the behaviour of an individual glacier. Fjord morphology controls the location and frequency of pinning points and the depth of water in which calving will occur (Figure 3.21). The behaviour of the individual glacier is also important. Some fjord glaciers are inherently unstable and prone to cyclic episodes of advance and decay, driven not by climate but by decoupling of the glacier from the stabilising effect of moraine banks or pinning points (Box 11.4). As a result, facies architecture within fjords may be complex, as illustrated in

Figure 10.9. Here a relatively simple facies pattern produced by the retreat and subsequent advance of a glacier down a glaciomarine fjord under stable sea-level conditions. Initially the glacier margin is stabilised via a moraine bank or subaqueous outwash fan (see Section 11.2). On becoming uncoupled from this pinning point the

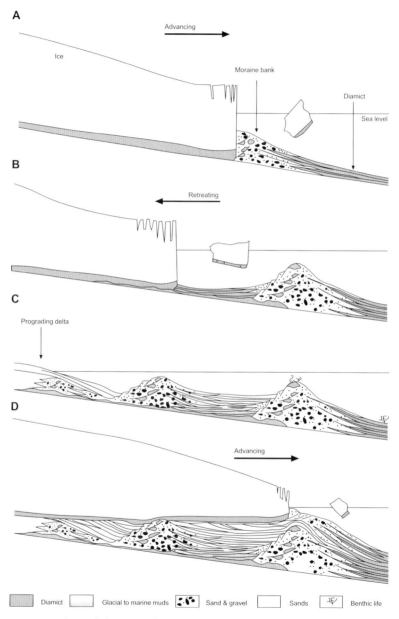

Figure 10.9   Hypothetical facies architecture associated with the retreat and advance of a glacier down a marine fjord. See text for facies descriptions. [Modified from: Powell (2003) in: *Glacial Landsystems* (ed. Evans), Arnold, figure 13.17, pp. 345–6]

11.1 GLACIOLACUSTRINE LANDFORMS

There is a wide variety of different types of glacial lake, although two broad types can be identified: those that form along an ice margin; and those that occur in a supraglacial setting. *Ice-marginal lakes* may form in front of glaciers or when ice dams water in a valley or against a hill side (see Figure 2.7). *Supraglacial lakes* can develop either where an ice-dammed lake expands over an ice margin (Figure 11.2) or in areas of complex ice-cored topography. The geomorphological products of these two categories of lake are different and will be considered in turn.

Figure 11.2 Supraglacial lake submerging part of an ice margin in Greenland. Note the debris bands on the glacier surface formed by ice-marginal thrusting. [Photograph: N.F. Glasser].

There are two types of glacier-lake interface at ice-marginal lakes; those at which large fans or deltas form, separating the ice margin from the lake, and those at which the ice front is actively calving (Figure 11.3). At calving margins small fans and moraine banks tend to form depending on whether the ice margin is stationary and upon the amount of meltwater discharge. The development of large deltas is, for example, favoured by a stable ice margin with a high meltwater–sediment discharge. The landforms associated with these different lake margins and with lakes in general are reviewed below.

A: Calving margin

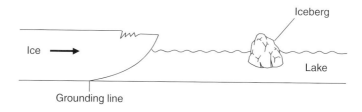

B: Ice-contact delta

Figure 11.3   Type of glacier–lake interface. (A) Calving margin. (B) Ice-contact delta-type margin.

1. **Glaciolacustrine landforms at non-calving ice margins.** At a stationary ice margin with high meltwater discharge an ice-contact delta may develop. Where sediment is delivered from a single or narrowly confined group of channels, delta fronts are arcuate in planform. In contrast, where meltwater streams switch from one side of a valley to another the delta front may be more linear in planform following the outline of the ice margin. The ice-contact face of the delta may be deformed and raised by post-depositional ice pushing and this ice-contact face is frequently subject to collapse and slumping when the ice support is removed. A *delta moraine* may develop where an ice front remains stationary for a considerable period of time. A delta moraine consists of a series of deltas or fans that coalesce along an ice margin to form a continuous ridge. The moraine is not formed from one but from many fans and deltas. Probably the longest delta moraine exposed on land is the Salpausselkä Moraine in Finland, which is over 600 km long and formed in the Baltic ice lake (Box 11.1).

2. **Glaciolacustrine landforms at calving ice margins.** Moraine ridges may develop at the grounding line. This is the line at which the ice begins to float and loses contact with its bed. These moraines are frequently referred to as De Geer moraines, washboard, or cross-valley moraines. They are low (<5 m), often asymmetric, ridges that are either straight or slightly concave in planform. This planform reflects the linear or concave morphology typical of a calving glacier margin in plan. They appear to form at the grounding line, which may be some

BOX 11.1: THE DELTA MORAINES OF THE BALTIC ICE LAKE

At the close of the last glacial cycle a very large lake lay along the southern margin of the Scandinavian Ice Sheet across what is the Baltic Sea today. Large delta moraines accumulated along the southern margin of the ice sheet. These are best developed in southern Finland where the Salpausselkä moraines occur. The moraines extend for over 600 km and are composed for the most part of outwash sediments deposited along a stationary ice margin. Fyfe (1990) studied the morphology and sedimentology of one of the three Salpausselkä moraines. She found that it was composed along its length of three geomorphological components.

1. Large individual deltas with braided tops, ice-contact deltas, which built up to the water level and were produced in relatively shallow water where the meltwater outflow was concentrated into a small number of concentrated outflows or portals.
2. Low narrow fans, grounding-line fans, which coalesce along the former ice margin and were formed in deeper water where the meltwater outflow was more distributed along the ice margin.
3. Small laterally overlapping subaqueous fans formed where the location of the meltwater portals was unstable due to calving into deep water. Consequently a large number of small fans developed along the margin.

The water depth and the nature of the subglacial drainage system – concentrated, distributed, stable or unstable – controlled the nature of the outwash accumulation that formed along the former ice margin, and therefore control the morphology of the delta moraine formed.

Source: Fyfe, G.J. (1990) The effect of water depth on ice-proximal glaciolacustrine sedimentation: Salpausselkä I, southern Finland. *Boreas*, **19**, 147–64.

distance behind the ice margin. Exceptionally fine examples have recently been revealed by sonar imagery from the German Bank area of the Scotian Shelf southwest of Nova Scotia (Box 11.2). There is considerable confusion surrounding the nomenclature and formation of these ridges. This in part reflects the fact that these ridges may form in several different ways (Figure 11.4); a concept known within geomorphology as *equifinality*. Some of these features form as seasonal push moraines produced by winter readvances of the grounding line in the same way as terrestrial push moraines (Figure 11.4A). It has been suggested that the calving of large icebergs may play a role in pushing up these ridges (see Figure 9.1B). Other examples may form as a consequence of the subglacial advection of deformation till to the ice margin and subsequent distal redistribution via a variety of sediment gravity flows. The internal structure of some larger ridges supports this model, with crude foreset beds of diamict dipping in a down-glacier direction

BOX 11.2: DE GEER MORAINES

The increasing availability of detailed three-dimensional bathymetry from continental shelves provides an opportunity to study the glacial landsystems formed by retreating tidewater glaciers. A particularly fine example is provided by Todd *et al.* (2007), who mapped an area of the Scotian Shelf revealing a bathymetric assemblage consisting of a series of superimposed landform suites. At the base is a series of subglacial flutes and drumlins, above which there are crevasse-squeeze ridges, De Geer moraines and larger moraine banks charting the pattern of ice-marginal retreat across the shelf. These landforms are in turn cross-cut by iceberg scour marks formed by the grounding of iceberg keels. Lindén and Möller (2005) investigated the internal sedimentary architecture of De Geer moraines in the coastal zone of northern Sweden, submerged during deglaciation by a freshwater lake occupying much of the Baltic Sea. They suggest that the proximal part of the moraines were built up in a submarginal position as stacked sequences of deforming bed diamicts, intercalated with glaciofluvial sediments formed in subglacial channels/canals, whereas the distal part of the ridges are formed from prograding layers of sediment deposited by sediment gravity flows moving forward of the grounding line. These sediment flows interfinger in a distal direction with glaciolacustrine sediments. In contrast, Blake (2000), working in northern Norway, emphasised the importance of subglacial and ice-marginal deformation of sediment deposited at the grounding line. No single model is likely to be appropriate at all localities due to the dynamic nature of a grounding line and its seasonal stability, which will be a function of ice velocity and the calving rate.

Sources: Blake, K.P. (2000) Common origin for De Geer moraines of variable composition in Raudvassdalen, northern Norway. *Journal of Quaternary Science*, **15**, 633–44. Lindén, M. and Möller, P. (2005) Marginal formation of De Geer moraines and their implications to the dynamics of grounding-line recession. *Journal of Quaternary Science*, **20**, 113–33. Todd, B.J., Valentine, P.C., Longva, O. and Shaw, J. (2007) Glacial landforms on German Bank, Scotian Shelf: Evidence for Late Wisconsinan ice-sheet dynamics and implications for the formation of De Geer moraines. *Boreas*, **36**, 148–69.

formed by the forward flow of the debris as it is released at the grounding line. Alternative explanations for De Geer moraines include the formation of ridges by the squeezing of saturated till into basal crevasses behind a calving ice margin. Similar ribbed moraines have also been described as forming well behind the ice margin in a subglacial position as a result of changes in the reheology of a deforming bed (see Section 3.33). Small sublacustrine fans can form at a steep calving ice margins wherever a meltwater portal exists. These small fans typically have steep gradients because deposition of the coarse fraction

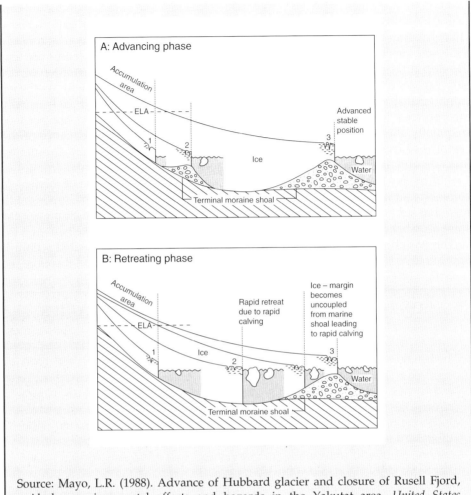

Source: Mayo, L.R. (1988). Advance of Hubbard glacier and closure of Rusell Fjord, Alaska—environmental effects and hazards in the Yakutat area. *United States Geological Survey Circular*, **1016**. [Modified from: Warren (1992) *Progress in Physical Geography* **16**, figure 11.4, p. 265]

morphology of the debris plume (Figure 11.5). At low water discharges the plume rises rapidly to the surface. Coarse tractional deposits are dumped close to the ice margin at the meltwater portal. Most of the fan is deposited on the distal slope by settling from the debris plume and by debris slumping or sliding down the distal face. This slumping gives rise to a series of foresets. At higher water and sediment discharges the plume remains in contact with the fan for longer and flattens or planes off the apex of the fan. Coarse sediment is also deposited at this point in sheets, and as scour and fill structures. Deposition by settling and by sediment gravity flows again builds up the distal part of the fan. At very high discharges deposition from the plume while it is in contact with the fan may deposit a barchanoid bar at the point at which the plume is detached from the fan (Figure 11.5).

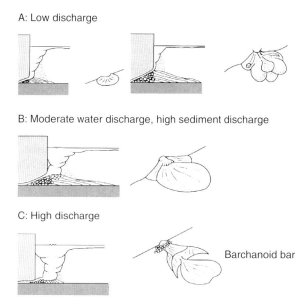

A: Low discharge

B: Moderate water discharge, high sediment discharge

C: High discharge

Barchanoid bar

Figure 11.5 Schematic representation of the variation in the discharge plume in front of a calving glacier in relation to water discharge and the impact of this variation on the morphology of the grounding line fan produced. (A) At low water discharges the fan grows progressively with coarse traction deposits dumped at the meltwater portal and by deposition as a series of large foresets on the distal flank from settling, slumping/sliding and sediment gravity flows. (B) At moderate sediment discharges the base of the sediment plume remains in contact with the fan longer. As a consequence a series of sheet-like deposits of sand and gravel with crude scour and fill structures is deposited on the top of the fan. Settling and sediment gravity flow deposits make up more distal slopes of the fan. (C) At high water discharges the plume remains in contact with the fan much longer and may deposit a migrating barchanoid bar along line of detachment between the plume and the sea bed. [Reproduced from: Powell (1990) in *Glacimarine Environments: Processes and Sediments* (eds J.A. Dowdeswell and J.D. Scourse), Geological Society Special Publication No. 53, figure 11.5, p. 58. Copyright © 1990, Geological Society Publishing House].

The morphology of grounding line fans is also controlled by the stability of the ice margin and the rate of discharge. At advancing ice margins grounding line fans will be small and heavily tectonised. In contrast, at retreating ice margins the fan morphology will depend upon: (i) the rate of ice-marginal retreat; (ii) the stability of the position of the meltwater portal; and (iii) the rate of the sediment discharge. Given these variables the following morphologies may develop.

1. **Advancing ice margin.** In this situation a small fan would develop in front of the ice margin during the summer ablation season when calving is high. These fans will be overrun and tectonised each winter when the glacier advances more rapidly due to lower calving rates.
2. **Stationary ice margin.** In this situation a single large fan will develop. If the rate of sedimentation is high this fan may grow to sea level and form an ice-contact delta.

formed along the margin of a fjord by waves generated by glacier calving. Iceberg plough marks are discussed in detail in papers by Reimnitz and Barnes (1974), Belderson *et al.* (1973), Kovacs and Mellor (1974), and Dowdeswell *et al.* (1993). Boulder pavements are discussed in detail by both Hansom (1983).

Andrews, J.T. (1963a) Cross-valley moraines of the Rimrock and Isortoq River valleys, Baffin Island, North West Territories. *Geographical Bulletin*, **19**, 49–77.

Andrews, J.T. (1963b) Cross-valley moraines of north-central Baffin Island: a quantitative analysis. *Geographical Bulletin*, **20**, 82–129.

Andrews, J.T. and Smithson, B.B. (1966) Till fabric of the cross-valley moraines of north-central Baffin Island, North West Territories, Canada. *Bulletin of the Geological Society of America*, **77**, 271–90.

Barnett, D.M. and Holdsworth, G. (1974) Origin, morphology, and chronology of sublacustrine moraines, Generator Lake Baffin Island, Northwest Territories, Canada. *Canadian Journal of Earth Science*, **11**, 380–408.

Beaudry, L.M. and Prichonnet, G. (1991) Late glacial De Geer moraines with glaciofluvial sediment in the Chapais area, Québec (Canada). *Boreas*, **20**, 377–94.

Belderson, R.H., Kenyon, N.H. and Wilson, J.B. (1973) Iceberg plough marks in the northeast Atlantic. *Palaeogeography, Palaeoclimatology, Palaeoecology*, **13**, 215–24.

Benn, D.I. (1988) Controls on sedimentation in a Late Devensian ice-dammed lake, Achnasheen, Scotland. *Boreas*, **18**, 31–42.

Boulton, G.S. (1986) Push-moraines and glacier-contact fans in marine and terrestrial environments. *Sedimentology*, **33**, 677–98.

Boulton, G.S. (1990) Sedimentary and sea level changes during glacial cycles and their control on glacimarine facies architecture, in *Glacimarine Environments: Processes and Sediments* (eds J.A. Dowdeswell and J.D. Scourse), Geological Society Special Publication No. 53, Bath, pp. 15–52.

Clemmensen, L.B. and Houmark-Nielsen, M. (1981) Sedimentary features of a Weishselian glaciolacustrine delta. *Boreas*, **10**, 229–45.

Dowdeswell, J.A., Villinger, H., Whittington, R.J. and Marienfield, P. (1993) Iceberg scouring in Scoresby Sund and on the East Greenland continental shelf. *Marine Geology*, **111**, 37–53.

Dowdeswell, J.A., Ó Cofaigh, C., Noormets, R., *et al.* (2008) A major trough-mouth fan on the continental margin of the Bellingshausen Sea, West Antarctica: The Belgica Fan. *Marine Geology*, **252**, 129–40.

Eyles, C.H. (1988) A model for striated boulder pavement formation on glaciated shallow marine shelves: an example from the Yakataga Formation Alaska. *Journal of Sedimentary Petrology*, **58**, 62–71.

Eyles, C.H. (1994) Striated boulder pavements. *Sedimentology*, **88**, 161–73.

Fyfe, G.J. (1990) The effect of water depth on ice-proximal glaciolacustrine sedimentation: Salpausselkä I, southern Finland. *Boreas*, **19**, 147–64.

Gore, D.B. (1992) Ice-damming and fluvial erosion in the Vestfold Hills, East Antarctica. *Antarctic Science*, **4**, 227–34.

Hansom, J.D. (1983) Ice-formed intertidal boulder pavements in the sub-Antarctic. *Journal of Sedimentary Petrology*, **53**, 135–45.

Johnson, P.G. and Kasper, J.N. (1992) The development of an ice-dammed lake: the contemporary and older sedimentary record. *Arctic and Alpine Research*, **24**, 304–13.

King, L.H. (1994) Till in the marine environment. *Journal of Quaternary Science*, **8**, 347–58.

Kovacs, A. and Mellor, M. (1974) Sea ice morphology and ice as a geologic agent in the southern Beaufort Sea, in The Coast and Shelf of the Beaufort Sea (eds J.C. Reed and J.E. Slater), Arctic Institute of North America, Arlington, VA, pp. 113–64.

Larsen, E., Longva, O. and Follestad, B.A. (1991) Formation of De Geer moraines and implications for deglaciation dynamics. *Journal of Quaternary Science*, **6**, 263–77.

Matthews, J.A., Dawson, A.G. and Shakesby, A. (1986) Lake shoreline development, frost weathering and rock platform erosion in an alpine periglacial environment, Jotunheimen, southern Norway. *Boreas*, **15**, 33–50.

McCabe, A.M. and Eyles, N. (1988) Sedimentology of an ice-contact glaciomarine delta, Carey Valley, Northern Ireland. *Sedimentary Geology*, **59**, 1–14.

Nielsen, N. (1991) A boulder beach formed by waves from a calving glacier; Eqip Sermia, West Greenland. *Boreas*, **21**, 159–68.

Ó Cofaigh, C., Dowdeswell, J.A., Allen, C.S., *et al.* (2005) Flow dynamics and till genesis associated with a marine-based Antarctic palaeo-ice stream. *Quaternary Science Reviews*, **24**, 709–40.

Ottesen, D., Dowdeswell, J.A., Benn, D.I., *et al.* (2008) Submarine landforms characteristic of glacier surges in two Spitsbergen fjords. *Quaternary Science Reviews*, **27**, 1583–99.

Peacock, J.D. and Cornish, R. (1989). *Glen Roy Area: Field Guide*. Quaternary Research Association, Cambridge.

Powell, R.D. (1990) Glacimarine processes at grounding-line fans and their growth to ice-contact deltas, in *Glacimarine Environments: Processes and Sediments* (eds J.A. Dowdeswell and J.D. Scourse), Geological Society Special Publication No. 53, Bath, pp. 53–73.

Powell, R.D. and Molnia, B.F. (1989) Glacimarine sedimentary processes, facies and morphology of the south-southeast Alaska shelf and fjords. *Marine Geology*, **85**, 359–90.

Reimnitz, E. and Barnes, P.W. (1974) Sea ice as a geologic agent on the Beaufort Sea Shelf of Alaska, in *The Coast and Shelf of the Beaufort Sea* (eds J.C. Reed and J.E. Slater), Arctic Institute of North America, Arlington, VA, pp. 301–53.

Sissons, J.B. (1977) Former ice-dammed lakes in Glen Moriston, Inverness-shire. *Transactions of the Institute of British Geographers*, **2**, 224–42.

Sissons, J.B. (1978) The parallel roads of Glen Roy and adjacent glens, Scotland. *Boreas*, **7**, 183–244.

Sissons, J.B. (1982) A former ice-dammed lake and associated glacier limits in the Achnasheen area, central Ross-shire. *Transactions of the Institute of British Geographers*, **7**, 98–116.

Stoker, M.S. and Holmes, R. (1991) Submarine end-moraines as indicators of Pleistocene ice-limits off northwest Britain. *Journal of the Geological Society, London*, **148**, 431–4.

Thomas, G.S.P. (1984) A Late Devensian glaciolacustrine fan-delta at Rhosemor, Clwyd, North Wales. *Geological Journal*, **19**, 125–41.

Zilliacus, H. (1989) Genesis of De Geer moraines in Finland. *Sedimentary Geology*, **62**, 309–17.

12: Palaeoglaciology

The aim of this chapter is to explain the science of *palaeoglaciology* – the methods by which former glaciers and ice sheets can be reconstructed from the landforms and sediments that they leave behind. Drawing on the processes, landforms and sediments of glacial erosion and deposition outlined in Chapters 5–11, we outline how these can be used to reconstruct former ice sheets. Examples are provided from the former mid-latitude ice sheets, including the North American (Laurentide), Scandinavian (Fennoscandian) and British–Irish ice sheets.

Fundamental to palaeoglaciology is the *glacial inversion method* (Figure 12.1), which is the conceptual process used to determine ice sheet evolution through time using the landforms exposed on the beds of former ice sheets. This process rests on the assumption that once we understand how certain landforms are created by ice sheets then we can use the distribution of these landforms to reconstruct their vertical and horizontal extent. By including other proxy data, such as chronological control by dating selected landforms or sediments, we can build a reconstruction that can be linked to relative sea-level history or linked to ice-core and marine records of climate change to model the behaviour of former ice sheets and their interaction with climate. Ice sheets respond dynamically to climate change over glacial cycles through changes in their extent and spatial configuration, and the landforms and sediments left behind after they have melted provide a critical record of this behaviour. It also provides a fitting conclusion to this book, since it integrates knowledge from throughout the book.

12.1 THE METHODS USED IN PALAEOGLACIOLOGY

A major impetus for the science of palaeoglaciology was the advent of optical satellite imagery in the late 1980s and early 1990s. Satellite images are useful because a single image covers a large area (100s of km^2) of the Earth's surface, making it possible to gain a coherent view of regional landform distributions. Mapping of the spatial distribution of landforms visible on satellite images is

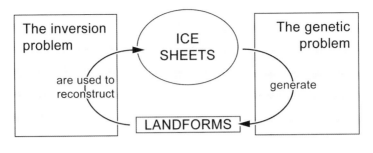

OUTPUTS

Former spatial extent
Former vertical extent
Former patterns of ice discharge
Palaeo-ice streams
Former ice divides and basal
thermal regime
History of growth and decay
Contribution to global sea level

Figure 12.1　The genetic and inversion problems in glacial geomorphology. [Modified from: Kleman *et al.* (2006), in *Glacier Science and Environmental Change* (ed P.G. Knight), Blackwell, plate 38.4, p. 197]

fundamental to palaeoglaciology. A basic outline of the steps required for a palaeo-glaciological reconstruction follows.

1. Obtain satellite image coverage of the area in question. Satellite images need to be acquired in the visible spectrum (optical images) and at a spatial resolution (say 30 m or less) so that landforms can be identified and mapped. Satellite images are often draped over a digital terrain model (DTM) to aid landform interpretation in areas of complex terrain. Aerial photographs might also be used to provide specific information on small areas where finer resolution is required.

2. Landform mapping. Mapped landforms are identified using established identification criteria (Table 12.1). Their location and boundaries are digitised on-screen and stored in a Geographical Information System (GIS) for later retrieval. If possible, maps should be checked in the field to ensure that landforms have been correctly identified. If it is available, information about striae and till stratigraphy can also be added at this stage. The most commonly used landforms in palaeoglaciological reconstructions are *ice-moulded subglacial landforms* (see Section 9.2.1), sometimes referred to as *streamlined subglacial landforms* (e.g., drumlins, flutes, megaflutes, megascale glacial lineations). To these information on ice-marginal positions and subglacial hydrology can be added by mapping terminal moraines, eskers and meltwater channels, although other landforms can also be used (see Section 12.2).

Table 12.1 Palaeoglaciological variables and the primary landforms that can be used to reconstruct them

Palaeoglaciological reconstruction	Landforms	Information provided
Former spatial extent of an ice sheet and its variation through time (i.e. location of the ice margin and how this fluctuated)	Terminal moraines Ice-marginal meltwater channels Eskers	Terminal moraines mark maximum former extent. Can be dated with ^{14}C, OSL or cosmogenic isotopes. Marginal meltwater channels indicate position of former ice margin and its recessional history. Eskers generally form at or close to the former ice margin, parallel or subparallel to ice-flow direction.
Former vertical extent of an ice sheet and its variation through time (i.e. former ice-surface elevation, ice thickness and how this changed through time)	Trimlines (see Box 12.5) Erratic boulders and erratics Ice-marginal meltwater channels Raised shorelines and raised beaches	Trimlines can be used to infer former vertical height. Erratic boulders deposited on mountains around the ice sheet indicate former debris transport paths. Elevation of erratic boulders on mountains provides an estimate of former vertical ice-sheet height. Marginal meltwater channels indicate former vertical dimensions of an ice sheet. Former ice thickness can be calculated from pattern of isostatic rebound following deglaciation.
Former patterns of ice discharge in an ice sheet (i.e. locations of major outlet glaciers and ice streams, former ice-flow trajectories)	Glacial lineations Striations Subglacial meltwater channels Eskers Erratic dispersal	Ice flow always parallel to orientation of glacial lineations. Can be used to identify location of former ice streams (see Box 12.4). Ice flow always parallel to orientation of striations. Subglacial meltwater channels generally parallel or subparallel to former ice-flow direction. Eskers generally form parallel or subparallel to former ice-flow direction. Patterns of erratic dispersal indicate former ice-flow trajectories.

Table 12.1   Continued.

Palaeoglaciological reconstruction	Landforms	Information provided
Former locations of ice divides and frozen-bed areas	Preglacial and periglacial landforms, e.g., rounded summits, fluvial valleys, cryoplanation terraces and pediments, and the presence of tors, blockfields and saprolites	Landform preservation provides information on former subglacial thermal regimes, that is, location of former frozen-bed areas and their spatial extent.
		Patterns of ice-divide migration can be traced through deglaciation. Long-lived former ice-divide locations may be indicated by areas of maximum isotatic rebound.
Overall history of ice-sheet growth and decay through time	Spatial distribution of all the above landforms and their relative ages	Landforms can be combined to make a full palaeoglaciological reconstruction.
		Can incorporate or make comparisons with results of numerical ice-sheet models or isostatic rebound models (see Box 12.6).
		Can incorporate or make comparisons with other proxy measures of global ice volume (e.g., oxygen isotope records).

3. **Data reduction.** The pattern of mapped landforms is simplified into a number of coherent landform systems using a *glacial inversion model* (Box 12.1). This step is required because across an ice sheet bed there may be many hundreds or even thousands of individual landforms, as well as striae and other point data (e.g., till-fabric analyses, dated landform surfaces). The most commonly used method is to convert landforms into a number of coherent '*flow sets*' (also called 'flow packages', 'fans' or 'swarms'), where each flow set reflects an ice-flow system that is spatially and temporally distinct (Figure 12.2). Established criteria are used to group landforms into these flow sets, including parallel concordance (i.e. similar orientation); proximity to one another; and morphometry (e.g., similar length, width, height, elongation ratios for streamlined features such as drumlins). It is also important to ensure that these flow sets, once defined, form a glaciologically plausible pattern.

4. The final step is the *palaeoglaciological reconstruction* itself. This involves making an overall interpretation of the patterns of events and their relative timing based on the patterns of mapped landforms and flow sets. It is also possible to combine

BOX 12.1: GLACIAL GEOLOGICAL INVERSION MODELS

In a series of papers, Johan Kleman and his co-workers have outlined the procedures used in glacial geological inversion models (i.e. a theoretical model that formalises the procedure of using the landform record to recon-struct ice sheets) to reconstruct past ice sheets from the glacial geomorpholo-gical record (Kleman and Borgström, 1996; Kleman *et al.*, 1997, 2006). Their inversion model comprises a classification system for glacial landform assem-blages and uses the following 'rules':

1. The basic control on landform creation, preservation and destruction is the location of the phase boundary between water and ice at or under the ice sheet base, that is the basal temperature, because this separates frozen from thawed material.
2. Basal sliding requires a thawed bed.
3. Glacial lineations can form only if basal sliding occurs.
4. Glacial lineations are formed parallel to local ice-flow directions and per-pendicular to the ice sheet surface contours at the time of creation.
5. Frozen-bed conditions inhibit the reshaping of the subglacial landscape.
6. Regional deglaciation is always accompanied by the creation of a spatially coherent but metachronous system of meltwater features, such as meltwater channels, eskers and glacial lake traces.
7. Eskers are formed in an inward-transgressive fashion inside a retreating ice front.
8. Meltwater channels will form the major landform record during frozen-bed deglaciations, whereas eskers are typically lacking under these conditions.

Using these assumptions, Kleman *et al.* (1997) mapped the landforms on the bed of the former Fennoscandian Ice Sheet and its change through time. They successfully combined their palaeoglaciological reconstructions with stratigra-phical records (e.g., till stratigraphy, cross-cutting striations) and dating chron-ologies (e.g., ^{14}C dates, to make time-dependent reconstructions of the Fennoscandian Ice Sheet through its growth and decay. The figure below shows their reconstructions of the Fennoscandian Ice Sheet at six time periods during the last glacial cycle. The reconstructions show the ice sheet extent, dispersal-centre locations (D) and inferred ice-flow patterns (arrows). Reconstructed parameters include the configuration of the ice sheet and its change over time, including the location of the main ice divides and the main ice-flow directions. Based on the preservation of areas of pre-Late Weichselian landscapes Kleman *et al.* (1997) inferred that the ice sheet had a frozen-bed core, which was only partly diminished in size by inward-transgressive wet-bed zones during the decay phase.

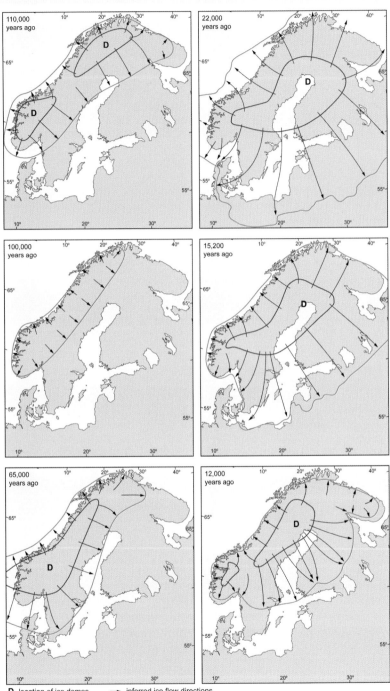

D location of ice domes   → inferred ice flow directions

Sources: Kleman, J. and Borgström, I. (1996) Reconstruction of palaeo-ice sheets: The use of geomorphological data. *Earth Surface Processes and Landforms*, **21**, 893–909. Kleman, J., Hättestrand, C., Borgström, I. and Stroeven, A. (1997) Fennoscandian palaeoglaciology reconstructed using a glacial geological inversion model. *Journal of Glaciology*, **144**, 283–99. Kleman, J., Hättestrand, C., Stroeven, A.P. *et al.* (2006) Reconstruction of palaeo-ice sheets – inversion of their glacial geomorphological record, in *Glacier Science and Environmental Change* (ed. P.G. Knight), Blackwell, Oxford, pp. 192–8. [Modified from: Kleman *et al.* (1997) *Journal of Glaciology*, **144**, figure 11, p. 297]

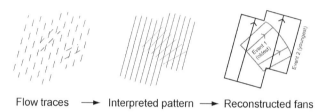

Flow traces → Interpreted pattern → Reconstructed fans

Figure 12.2 Example of how ice-flow systems can be reconstructed from landform evidence. In this example, two events are recognised on the basis of the flow traces identified. The landforms relating to Event 2 are interpreted as the youngest because they overprint those of Event 1.

these data with existing stratigraphical records and dating techniques to produce a more comprehensive picture of ice-sheet growth and decay (Box 12.1). The mapped landforms and the flow sets they define are then used to reconstruct glaciological parameters such as the location of major outlet glaciers and ice streams, the directions of former ice flow, the location of ice divides, the position of glacier termini, changes in basal thermal regime and the patterns of ice recession.

12.2 THE KEY LANDFORMS USED IN PALAEOGLACIOLOGY

A number of different landforms are used in palaeoglaciological reconstructions (Table 12.1). The following eight groups of landforms provide the most information about former subglacial conditions and ice-sheet dynamics (see Chapters 9–11). The most powerful palaeoglaciological reconstructions, however, are those that use a combination of many different landforms.

1. *Ice-moulded* or *streamlined subglacial landforms* (Box 12.2) cut in bedrock (*grooves* and *mega-grooves*) or made of sediment (e.g., *glacial lineations*, *flutes* and

megaflutes, megascale glacial lineations and drumlins) are formed parallel to ice-flow direction (Figure 12.3). Their precise origin may not always be clear, but we do know that all these features form in a subglacial position (at the ice–sediment/bedrock interface) when the ice-sheet bed is thawed and where sufficient subglacial debris is available (Figure 12.4). These streamlined landforms are found in both terrestrial and marine settings, where they record fast glacier flow and, in some cases, the location of terrestrial and marine-based palaeo-ice streams. All these features provide important information on former ice-flow directions and their change through time.

BOX 12.2: THE ORIGIN AND SIGNIFICANCE OF STREAMLINED GLACIAL LANDFORMS

Some of the most commonly used landforms in ice-sheet reconstructions are a family of ice-moulded and streamlined landforms comprising glacial lineations, drumlins, megaflutes and megascale glacial lineations (MSGL). These landforms are visible on satellite images as large-scale patterns of topographic streamlining, sometimes with distinctive cross-cutting relationships. The image below shows a Landsat subscene (left panel) and interpretation (right panel) of the glacial geomorphology of the area around Laguna Cabeza del Mar (near Seno Otway in southern Patagonia). At least two sets of streamlined glacial lineations can be traced to former ice-marginal positions in the northeast of the image. The former ice margins are marked by moraines, sandar and meltwater channels. Understanding the genesis and significance of these landforms is clearly crucial to all palaeoglaciological reconstructions. Clark (1993, 1994) provided a glaciological explanation for the origin of these streamlined landforms based on the premise that streamlined landforms are created subglacially. Glacial lineations form parallel to ice-flow direction, providing a 'marker' that records deformation during ice-flow events. Clark argues that the palaeoglaciological record provides the surface or plan expression of subglacial deformation. This work suggests that MSGL (up to 20 km in length and with length : width ratios of up to 50 : 1) were formed under conditions of fast ice flow and their presence may thus record the locations of former ice streams or surge events (Stokes and Clark, 2002). Extensive sets of lineations must have been formed approximately synchronously, thereby indicating that lineation generation occurs over a wide range of glaciodynamic conditions, from submarginal positions to the interior portions of ice sheets. These streamlined glacial landforms are now commonly used in palaeoglaciological reconstructions. For example, using satellite imagery, Stokes and Clark (2002) mapped over 8000 glacial lineations associated with an ice stream of the Laurentide Ice Sheet near Dubawnt Lake, District of Keewatin, Canada and used these to reconstruct areas of former high ice-velocity under the Laurentide Ice Sheet.

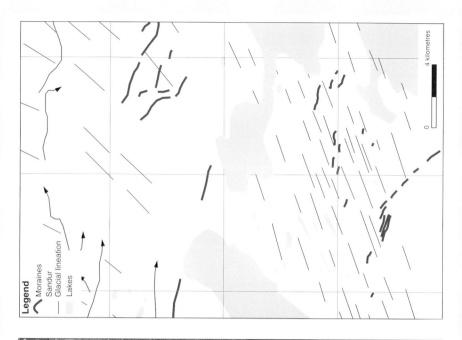

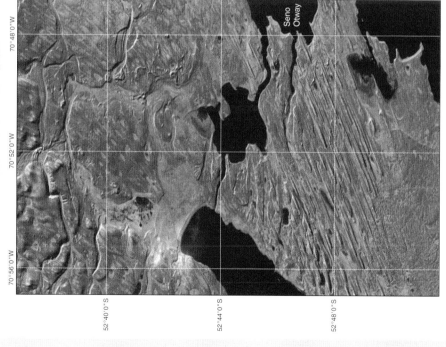

Sources: Clark, C.D. (1993) Mega-scale glacial lineations and cross-cutting ice-flow landforms. *Earth Surface Processes and Landforms*, **18**, 1–29. Clark, C.D. (1994) Large-scale ice-moulding: A discussion of genesis and glaciological significance. *Sedimentary Geology*, **91**, 253–68. Stokes, C.R. and Clark, C.D. (2002) Are long sub-glacial bedforms indicative of fast ice flow? *Boreas*, **31**, 239–49. [Modified from: Glasser *et al.* (2008) *Quaternary Science Reviews*, **27**, figure 2, p. 371]

Figure 12.3   Rock-cored drumlin to the east of the Southern Alps of New Zealand. Former ice flow was right to left. There is a person in the left-centre of the photograph for scale. [Photograph: N.F. Glasser]

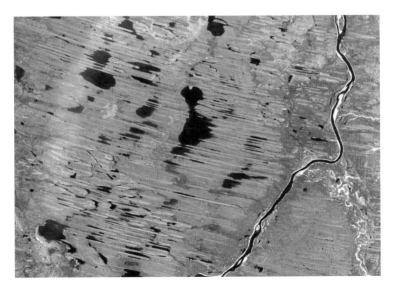

Figure 12.4   Satellite image of megascale glacial lineations (MSGL) formed under the Laurentide Ice Sheet in Canada. Former ice flow was from right to left. [Image courtesy: C.D. Clark]

2. *Non-ice-moulded subglacial landforms*, including *geometrical ridge networks* and *crevasse-squeeze ridges*. These features are believed to form by the squeezing of basal till into subglacial crevasses, and are commonly associated with surge-type behaviour.

3. *Ribbed (Rogen) moraines*, which are subglacial landforms comprising discontinuous, subparallel to parallel ridges oriented transverse to former ice flow. They are typically steep-sided, regularly spaced and made of diamict, sand or gravel. They have a variety of morphological forms and their precise mode of formation is debated. In some cases it can be demonstrated that they have been reshaped parallel to ice flow by ice overriding, indicating sediment deformation. In this case they can be regarded as ice-moulded subglacial landforms. It has also been argued that ribbed moraines reflect large-scale sediment deformation during inward-transgressive subglacial thawing at the ice-sheet bed.

4. *Ice-marginal landforms*, including *ice-marginal moraines*, which appear on satellite images as prominent cross-valley single moraines, in more complicated moraine systems with multiple ridges or as areas of hummocky moraine. They can be linear, curved, sinuous or saw-toothed in plan. They are often associated with other ice-marginal features such as *meltwater channels, ice-contact fans, kames, kame terraces, kettle holes* and *outwash plains* (*sandur*). Ice-marginal landforms generally reflect periods when the ice margin was stationary for some length of time. In marine or lacustrine settings, non-ice-moulded subglacial landforms include regularly spaced *transverse ridges, De Geer moraines, delta moraines, moraine banks, grounding-line fans, trough-mouth fans* and *push moraines* formed during ice retreat.

5. *Eskers*, which are sinuous ridges of sand and gravel, sometimes braided or beaded in planform, with surface pits and mounds. Most eskers form in ice-walled tunnels where sediment is transported and deposited by englacial or subglacial meltwater. They tend to be orientated subparallel or parallel to the former ice-flow direction and often run down-valley. They indicate warm-based ice sheet conditions and form time-transgressively close to the ice margin during ice-sheet recession phases.

6. *Meltwater channels*, which appear as incised linear features with abrupt inception and termination. They often contain no contemporary drainage and individual channels may either follow or cut across the local slope direction. Variants include *subglacial meltwater channels* cut beneath the ice and *ice-marginal meltwater channels*, cut adjacent to the ice margin, for example during incremental recession.

7. *Striae*, which are millimetre-scale wide but metre-scale long linear scratches on bedrock surfaces, and other *bedrock abrasional marks*. These microscale landforms can be identified only in the field, but they are invaluable because they provide important information about former ice-flow directions and basal thermal regime and how these varied through time.

8. *Unmodified preglacial landforms* often appear in the core areas of former ice sheets, especially in upland areas. These landforms include plateau surfaces sometimes with well-developed tors and associated saprolites, boulder fields and boulder

depressions, as well as fluvial valleys. These landforms are generally believed to develop by subaerial weathering and erosion processes. Their survival beneath an ice sheet indicates that former ice-sheet velocities were low and that the ice sheet was cold-based, at least for the majority of the time.

In areas lacking landform detail, *sedimentary evidence* may also be useful. Till-fabric analysis and till geochemical signatures may be used to infer former ice movement directions, as can directions of erratic dispersal.

12.3 FORMER SUBGLACIAL THERMAL REGIMES

Understanding former subglacial thermal regimes (see Section 3.4.1) is vital for an understanding of former ice sheets and the landforms they created (Figure 12.5). This is because the former subglacial thermal regime (warm-based *or* thawed-bed conditions versus cold-based *or* frozen-bed conditions) determines whether landforms are created (thawed bed) or preserved (frozen bed) beneath an ice sheet. One productive approach to this question is to compare the patterns of ice flow and subglacial thermal organisation of contemporary ice sheets with the patterns of ice flow indicated by landforms on the beds of former ice sheets such as the Laurentide or Fennoscandian Ice Sheets. By adopting this approach it is possible to develop conceptual models of the subglacial thermal organisation beneath ice sheets (Figure 12.6). Using the Antarctic Ice Sheet as an example, it appears that there are four major dynamic components of ice sheets that are important to understanding ice dynamics, landform creation and preservation beneath ice sheets. These are related directly to the spatial organisation of frozen- and thawed bed zones under the ice sheet.

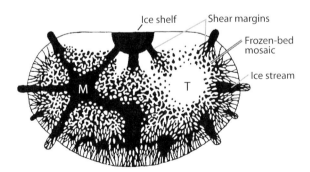

Figure 12.5 The range of possible subglacial thermal regimes in a former ice sheet. The figure shows a hypothetical ice sheet with a terrestrial dome (T), a marine dome (M), and an ice shelf. White areas indicate frozen-bed conditions and black areas indicate thawed-bed conditions. [Modified from: Kleman and Glasser (2007), *Quaternary Science Reviews*, **26**, Figure 1, p. 586]

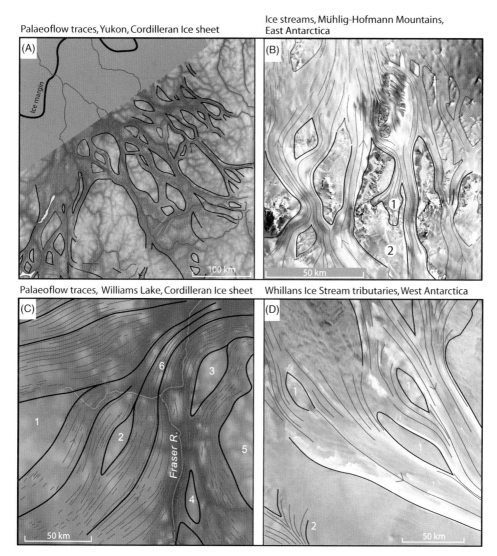

Palaeoflow traces, Yukon, Cordilleran Ice sheet

Ice streams, Mühlig-Hofmann Mountains, East Antarctica

Palaeoflow traces, Williams Lake, Cordilleran Ice sheet

Whillans Ice Stream tributaries, West Antarctica

Figure 12.6 Similarities between the frozen- and thawed-bed patterns in formerly glaciated areas (A and C) and the surface of the contemporary Antarctic Ice Sheet (B and D). (A) The organisation of flow traces (red lines) in the north-central Cordilleran Ice Sheet. (B) Ice streams passing through the Mühlig–Hofmann Mountains in East Antarctica. Black lines indicate the boundaries between ice-streams and low-velocity areas. Number 1 indicates an ice-stream tributary draining ice from a large frozen-bed patch (labelled '2'). (C) Flow traces south of Prince George, British Columbia, Canada. Symbols as in (A). Numbers 1–5 indicate inferred frozen-bed patches without flow traces comparable to the main flow pattern (6). (D) Major tributaries to the Whillans Ice Stream in West Antarctica. Number 1 indicates lenticular frozen-bed patches. Number 2 indicates the upper end of an ice-stream tributary. [Modified from: Kleman and Glasser (2007), *Quaternary Science Reviews*, **26**, figure 8, p. 590]

1. *Frozen-bed patches* are present under both contemporary and palaeo-ice sheets. In the contemporary Antarctic Ice Sheet, for example, frozen-bed patches occur on the ridges of higher land that lie between fast-flowing outlet glaciers and ice streams. Under former mid-latitude ice sheets the evidence for frozen-bed patches includes areas of the ice sheet bed that display a complete absence of glacial erosion or preserved landforms. Cosmogenic isotope dating has been used to confirm quantitatively the absence of erosion in many areas where preserved landforms are thought to represent relict surfaces. Often these preserved patches are adjacent to zones showing intense reshaping of the bed (e.g., the formation of glacial lineations) or glacial erosion (Box 12.3).

BOX 12.3: THE IMPORTANCE OF FORMER SUBGLACIAL THERMAL REGIME

One of the most important recent advances in palaeoglaciology is the realisation that ice sheets only destroy previous landforms and create new ones when they are thawed at the bed. This is because meltwater is required for subglacial sliding and sediment deformation to occur. Under frozen-bed conditions, the ice is frozen to its substrate and erosion is severely limited. Consequently geomorphologists have identified certain landform assemblages that are related to thawed-bed conditions, erosion and reshaping of the ice-sheet bed, and others to protection and non-erosion under frozen-bed conditions (Kleman, 1994; Kleman and Borgström, 2004). For example, Hättestrand and Stroeven (2002) studied the Parkajoki area in northeastern Sweden near the central area of the former *Fennoscandian Ice Sheet* (FIS). The landscape here is preglacial in its appearance and is dominated by landforms created by subaerial weathering and erosion processes, such as well-developed tors and associated saprolites, boulder fields and boulder depressions. The glacial geomorphology is dominated by lateral and proglacial meltwater channels. Glacial landforms indicating thawed-bed conditions and reshaping by glacier sliding (e.g., glacial lineations, drumlins, striae) are generally lacking. Because of its location, near the centre of the former FIS, these authors attributed landform preservation to frozen-bed conditions that protected the landscape from glacial erosion during recent glacial cycles. In a subsequent paper Stroeven *et al.* (2002) used cosmogenic ^{10}Be and ^{26}Al exposure age data from three tors and a meltwater channel in the Parkajoki area to test this hypothesis of landscape preservation through multiple glacial cycles. Their exposure age data for tors (exposure ages of between 79 and 37 ka in an area deglaciated at ~ 11 ka) suggest that the tors have indeed survived at least two complete glacial cycles. These data have now also been confirmed by other exposure age dating studies from large areas beneath the former FIS (e.g., Fabel *et al.*, 2002). Finally, there is also evidence in the form of the distribution of ribbed moraine that both the former FIS and the *Laurentide Ice Sheet* (LIS) were frozen-based during the Last Glacial Maximum

(Kleman and Hättestrand, 1999). These authors reasoned that the occurrence of ribbed moraines on the beds of the FIS and LIS can be used to determine the former spatial distribution of frozen- and thawed-bed conditions. They argued that ribbed moraines were formed by brittle fracture of subglacial sediments, induced by high stress at the boundary between frozen- and thawed-bed conditions resulting from the across-boundary difference in basal ice velocity. Together, all these studies indicate that an appreciation of the full range of former subglacial thermal regimes is vital to understanding the behaviour of former ice sheets and their impact on the landscape.

Sources: Fabel, D., Stroeven, A.P., Harbor, J., *et al*. (2002) Landscape preservation under Fennoscandian ice sheets determined from *in situ* produced ^{10}Be and ^{26}Al. *Earth and Planetary Science Letters*, **6266**, 1–10. Hättestrand, C. and Stroeven, A.P. (2002) A relic landscape in the central Fennoscandian glaciation: geomorphological evidence of minimal Quaternary glacial erosion. *Geomorphology*, **44**, 127–43. Kleman, J. (1994) Preservation of landforms under ice sheets and ice caps. *Geomorphology*, **9**, 19–32. Kleman, J. and Borgström, I. (1994) Glacial landforms indicative of a partly frozen bed. *Journal of glaciology*, **135**, 225–64. Kleman, J. and Hättestrand, C. (1994) Frozen-bed Fennoscandian and Laurentide ice sheets during the last glacial maximum. *Nature*, **402**, 63–6. Stroeven, A.P., Fabel, D., Hättestrand, C. and Harbor, J. (2002) A relict landscape in the centre of Fennoscandian glaciation: cosmogenic radionuclide evidence of tors preserved through multiple glacial cycles. *Geomorphology*, **44**, 145–54.

The boundaries between preserved patches (frozen-bed) and reshaping/erosion (thawed-bed) are often sharp, implying that the transition between the two is abrupt. Relict surfaces are interpreted to have been preserved because they were covered by an ice sheet that was dominated by frozen-bed conditions. Frozen-bed patches have now been identified at a wide range of spatial scales and in a range of paleoglaciological contexts, embracing large areas such as frozen-based interstream ridges in ice stream webs (scales of tens to hundreds of kilometres), right down to mosaics of small frozen patches in transition zones between frozen and thawed beds (scales of tens of metres).

2. Ice streams are narrow zones of fast flow in ice sheets that are otherwise dominated by slow sheet flow (Figure 12.5). They are defined as regions in a grounded ice sheet in which the ice flows much faster than in regions on either side. They exist as both 'pure' ice streams, where fast-flowing ice is surrounded on all sides by slow-flowing ice, and as topographically controlled ice streams, where their location is determined to an extent by the location of bedrock depressions or valleys. Typical ice-stream velocities are 0.5–1 km per year, up to two orders of magnitude faster than the surrounding sheet-flow areas. Geophysical studies of the contemporary Antarctic Ice Sheet show that ice streams can develop only where there is a thawed bed and a sufficient thickness of subglacial sediment to allow decoupling from the bed by high subglacial water pressures or subglacial deformation. Many of the contemporary Antarctic ice streams partly follow bedrock troughs where flow-parallel channelling generates the large quantities of

frictional heat required to sustain fast ice flow. Ice streams have also been recognised on the beds of former ice sheets. The locations of these *palaeo-ice streams* have been identified using a range of criteria (Box 12.4). There is evidence that palaeo-ice streams drained large areas of the former Laurentide, Fennoscandian, Cordilleran and British–Irish ice sheets (Figure 12.7). These ice streams would have been very important in determining the overall dynamics of these ice masses.

BOX 12.4: IDENTIFYING FORMER ICE STREAMS (PALAEO-ICE STREAMS)

Ice streams are critical components of both contemporary and palaeo-ice sheets. Using the characteristics of contemporary ice streams such as those in Antarctica, Stokes and Clark (1999, 2001) have proposed a set of criteria that can be used to identify the locations of ice streams on former ice-sheet beds as set out in the table below. In all, eight geomorphological criteria are identified as indicative of former ice-stream behaviour. It is of course highly unlikely that all eight of these criteria will be met in any one location, but collectively they provide a good set of objective criteria for identifying palaeo-ice streams.

Geomorphological criteria for identifying former ice streams.

Contemporary ice-stream characteristic	Geomorphological signature
Characteristic shape and dimensions	1. Characteristic shape and dimensions (>20 km wide and >150 km long)
	2. Highly convergent flow patterns
Rapid velocity	3. Highly attenuated subglacial landforms (length : width ratio > 10 : 1)
	4. Boothia-type erratic dispersal trains
Sharply delineated margins	5. Abrupt lateral margins
	6. Ice stream shear margin moraines
Deformable bed conditions	7. Glaciotectonic deformation of sediments or bedrock
Focused sediment delivery	8. Submarine sediment accumulations (e.g., 'trough-mouth fans' or 'till deltas' where marine-terminating)

Sources: Stokes, C.R. and Clark, C.D. (1999) Geomorphological criteria for identifying Pleistocene ice streams. *Annals of Glaciology*, **28**, 67–74. Stokes, C.R. and Clark, C.D. (2001) Palaeo-ice streams. *Quaternary Science Reviews*, **20**, 1437–57. [Table modified from: Stokes and Clark, (2001) *Quarternary Science Reviews*, Table 6, p. 145.]

---

3. *Ice-stream tributaries* flow at velocities intermediate between full ice-stream and sheet flow (Figure 12.6A). Tributaries link the large interstream areas (cold-based ice), to the major ice streams of the contemporary Antarctic Ice Sheet (Figure 12.6B). The role of these ice-stream tributaries in determining the overall

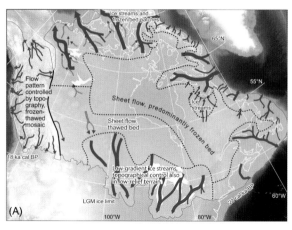
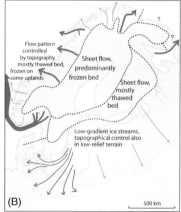

Figure 12.7 Reconstruction of former subglacial thermal organisation at the Last Glacial Maximum (LGM) or near-LGM for (A) the North American Ice Sheet and (B) the Fennoscandian Ice Sheet. In both cases the patterns have been reconstructed from the distribution of landforms mapped in remotely sensed images. [Modified from: Kleman and Glasser (2007), *Quaternary Science Reviews*, **26**, figure 13, p. 594]

BOX 12.5: THE VERTICAL DIMENSIONS OF FORMER ICE SHEETS

Establishing the former horizontal extent of a palaeo-ice sheet is relatively simple because this can be determined from the distribution of terminal moraines and associated subglacial landforms. Establishing the former vertical extent (surface height) of an ice sheet is, however, much more complicated. One approach that has now been widely adopted is to look for trimlines that mark the former vertical extent of an ice mass. Brook *et al*. (1996) pioneered this approach for the Fennoscandian ice sheet by obtaining cosmogenic isotope ages (^{10}Be and ^{26}Al) for bedrock and boulder samples both above and below a trimline that separates ice-scoured bedrock (below) from weathered bedrock (above) in the mountains of western Norway. The weathered blockfields on the mountain summits yielded an exposure age of > 55 ka, suggesting that this surface formed before the *Last Glacial Maximum* (LGM). Samples from the glaciated bedrock below the trimline yielded much younger exposure ages of 26–21 ka. Brook *et al*. (1996) were therefore able to reconstruct with some degree of confidence the former vertical extent and surface profile of the Last Glacial Maximum (LGM) ice sheet in this area. This approach has now been used to estimate the thickness of former ice sheets elsewhere, for example in Northwest Scotland (Stone *et al*., 1998). Other complementary techniques that can be used to investigate the former vertical dimensions of ice-sheets include joint-depth measurements and chemical analyses of weathering products in soils above and below trimlines

(Ballantyne, 1994; McCarroll *et al.*, 1995) and the degree of rock surface weathering (Nesje *et al.*, 2004). Cosmogenic isotope dating techniques have now also been used to establish the rates of vertical thinning of the contemporary Antarctic Ice Sheet since the LGM. By obtaining exposure ages for erratic rocks deposited on nunataks of the ice sheet it is possible to establish the pattern and rates of vertical ice-sheet retreat of the Antarctic Ice Sheet (Bentley *et al.*, 2006; Mackintosh *et al.*, 2007).

Sources: Ballantyne, C.K. (1994) Gibbsitic soils on former nunataks: Implications for ice sheet reconstruction. *Journal of Quaternary Science*, **9**, 73–80. Bentley, M.J. Fogwill, C.J. Kubik, P.W. and Sugden, D.E. (2006) Geomorphological evidence and cosmogenic ^{10}Be/^{26}Al exposure ages for the last glacial maximum and deglaciation of the Antarctic peninsula ice sheet. *GSA Bulletin*, **118**, 1149–59. Brook, E.J., Nesje, A., Lehman, S.J., *et al* (1996) Cosmogenic nuclide exposure ages along a vertical transect in western Norway; implications for the height of the Fennoscandian ice sheet. *Geology*, **24**, 207–10. Mackintosh, White, A., Fink, D., Gore, D., *et al* (2007) Exposure ages from mountain dipsticks in Mac. Robertson land, east Antarctica, indicate little change in ice sheet thickness since the last glacial maximum. *Geology*, **35**, 551–4. McCarroll, D., Ballantyne, C.K., Nesje, A. and Dahl, S.-O. (1995) Nunataks of the last ice sheet in northwest Scotland. *Boreas*, **24**, 305–23. Nesje, A., McCarroll, D. and Dahl, S.-O. (1994) Degree of rock surface weathering as an indicator of ice-sheet thickness along an east–west transect across southern Norway. *Journal of Quaternary Science*, **9**, 337–47. Stone, J.O., Ballantyne, C.K. and Fifield, L.K. (1998) Exposure dating and validation of periglacial weathering limits, northwest Scotland. *Geology*, **26**, 587–90.

BOX 12.6: NUMERICAL MODELLING OF FORMER ICE SHEETS: THE BRITISH ISLES ICE SHEET

High-resolution numerical computer models are an important tool in glacial geology because they can be used to make quantitative palaeoglaciological predictions about Quaternary ice sheets. Models can be used to predict parameters such as the thickness, distribution and age of tills, the distribution of indicator erratics, the distribution of eskers and tunnel valleys, relative sea-level change and the glacial sedimentary and isotopic flux to the oceans. Boulton and Hagdorn (2006) used a thermo-mechanically coupled numerical ice-sheet model, driven by a proxy climate, to explore the key properties of the last British Isles Ice Sheet – an ice sheet that had a relatively low surface profile, low summit elevation and extensive, elongated lobes at its margin. Their approach was to determine the boundary conditions that permit the simulated ice sheet to mimic the evolution of the real ice sheet through the last glacial cycle. Their simulations show how a British Isles Ice Sheet may have been confluent with a Scandinavian Ice Sheet during some parts of its history and how unforced periodic and asynchronous oscillations might occur in different parts of the ice sheet margins. Marginal ice lobes are a reflection of ice-stream development within the ice sheet. Such ice streams can be ephemeral, with 'dynamic ice streams' located because of ice sheet properties, or 'fixed ice streams', with their location determined by subglacial bed properties or topography. Boulton and Hagdorn (2006) concluded that the simulations that best satisfy the glacial geological constraints of ice-sheet extent, ice-surface elevation and relative sea levels are those with major fixed ice streams that strongly draw down the ice-sheet surface. In their simulations the core upland areas of the ice sheet were cold-based at the Last Glacial Maximum, ice-stream velocities were between 500 and 1000 m yr^{-1} (compared with velocities of 10–50 m yr^{-1} in interstream zones), shear stresses were as low as 15–25 kPa under the ice streams (compared with 70–110 kPa in upland areas) and 60–84% of the ice flux was delivered to the margin via ice streams. Shown below is output from the computer model Boulton and Hagdorn (2006).

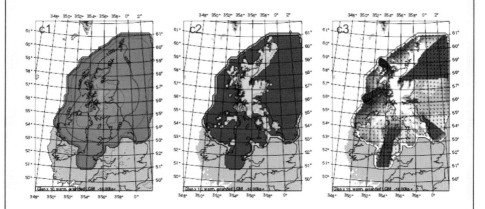

Output is for the Last Glacial Maximum (LGM) around 17 000 years ago. c1 shows the surface form of the ice sheet, c2 shows areas of basal melting (in red) and basal freezing (in blue), and c3 shows basal ice velocities, with red areas indicating high velocities and white areas indicating frozen-bed areas with zero velocities.

Source: Boulton, G.S. and Hagdorn, M. (2006) Glaciology of the British Isles ice sheet during the last glacial cycle: Form, flow, streams and lobes. *Quaternary Science Reviews*, **25**, 3359–90. [Modified from: Boulton and Hagdorn (2006) *Quaternary Science Reviews*, **25**, figure 14, p. 3382]

dynamics of large ice masses is still not fully understood but it is possible that the migration of tributaries is partly responsible for the switching on and off of ice streams.

4. *Lateral shear zones* develop at the contact between fast-flowing ice in ice streams or ice-stream tributaries and slow ice in interstream ridges. In the landform record, the boundaries of palaeo-ice streams, tens to hundreds of kilometres in length, have been interpreted as the sites of lateral shear zones. Sometimes, but not always, their location is marked by an *ice stream shear-margin moraine* (Figure 12.8). These features range in length from 11 to 22 km, with constant widths of around 500 m, and heights between 10 and 50 m. Examples exist of location control by topography and also of palaeoshear zones without any topographic or substratum control (Figure 12.6C). On the contemporary Antarctic Ice Sheet, lateral shear zones appear as linear and often intensely crevassed structures up to several hundred kilometres in length (Figure 12.6D). The occurrence of long, fully continuous shear zones without any obvious external control on location strongly suggests that internal differences

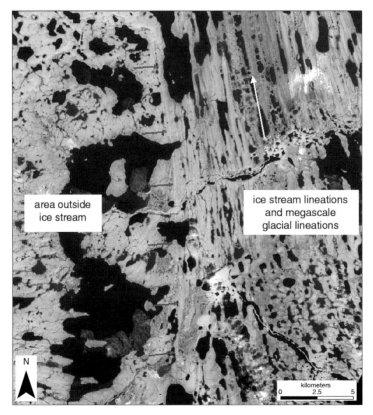

Figure 12.8   Landsat TM satellite image showing the location of ice stream shear-margin moraines (arrowed) on Storkerson Peninsula, Canada. [Image courtesy of: C.R. Stokes]

in material properties, for example shear-softened, warmer and more deformable ice in the shear margins, play an important role in determining the upstream, and possibly also downstream, propagation of fast ice flow.

Reconstructions of the basic arrangement of these ice sheet components for the Laurentide and Fennoscandian ice sheets show that these are similar in configuration to the contemporary Antarctic ice sheet. Both the Laurentide and Fennoscandian ice sheets had relatively slow-flowing interiors and were drained by large peripheral ice streams. Ice streams were arranged in a basically radial pattern, with individual ice streams separated from each other by topographically guided frozen-bed patches on intervening ridges (Figure 12.7).

12.4 PALAEOGLACIOLOGICAL RECONSTRUCTIONS

As an example of how these principles can be used to make a simple palaeoglaciological reconstruction we start with a map of the distribution of landforms on a hypothetical ice-sheet bed (Figure 12.9A) and outline how these landforms can be used to infer the basic shape and dimensions of a former ice sheet (Figure 12.9B).

1. **Terminal moraines.** Large terminal moraines mark the outer limit of the ice sheet. In some places the terminal moraine is a single ridge; in other places it may be a series of subparallel ridges. There may be large sandar (outwash plains) leading away from the terminal moraines wherever subglacial or supraglacial streams emerged from the ice margin. The terminal moraines can be used to reconstruct the size of the ice sheet at its maximum extent.

2. **Unmodified preglacial landscape.** In the core of the former ice sheet is a landscape dominated by the preservation of largely unmodified preglacial landforms on uplands. These landforms were created by subaerial weathering and erosion processes, and include well-developed tors and associated saprolites, boulder fields and boulder depressions. The preservation of these landforms is used to infer that this is the location of the main (first order) ice divide. Here former ice-sheet velocities were low and the ice sheet was cold-based, allowing preservation of components of the preglacial landscapes.

3. **Streamlined subglacial landforms.** Streamlined subglacial landforms are common on the former ice sheet bed. In the east, there are four separate systems of streamlined subglacial landforms, each representing a different flow set. Since these landforms are created parallel to former ice flow and because each set has a different orientation, we can infer that they indicate four distinct flow sets, each separated by subsidiary ice divides. In contrast, the streamlined subglacial landforms in the north and south have sharply delineated lateral margins. The maximum extent of the southern set is marked by large lobate moraine systems that project beyond the general extent of the ice-sheet terminal moraines. These south-flowing outlet glaciers would therefore be interpreted as the sites of palaeo-ice streams. There is no terminal moraine associated with the northward-flowing set

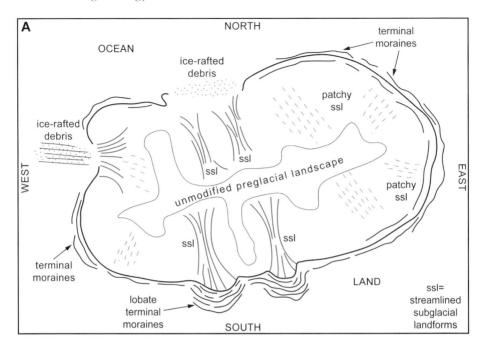

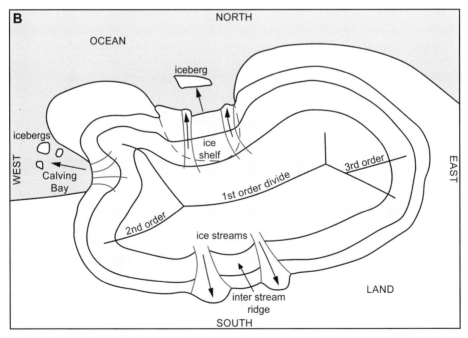

Figure 12.9 Example of a palaeoglaciological reconstruction. (A) Overview of mapped landform evidence. (B) Reconstruction of the former ice sheet based on the landform record. Note that the reconstruction assumes that the mapped glacial landforms formed more-or-less synchronously and near the time of the ice-sheet maximum, that is, there is little information about patterns of ice recession.

of streamlined subglacial landforms, but ice-rafted debris on the ocean floor indicates that there was a calving glacier or ice shelf here. These landforms could therefore be interpreted as the sites of palaeo-ice streams feeding into an ice shelf. On the western side of the ice sheet there is an area of streamlined subglacial landforms, this time without sharp lateral boundaries, no terminal moraine and with ice-rafted debris on the ocean floor. This again indicates a marine setting, but we cannot reconstruct a palaeo-ice stream in this location because the landforms lack the diagnostic abrupt lateral boundaries. Instead, the inference is that a fast-flowing tidewater glacier drained into a calving bay here.

This example is highly simplified and for scale reasons it omits many of the landforms listed in Section 12.3, which would also provide useful information about the former ice sheet. The 'real world' is likely to be much more complex for the following reasons.

1. We have illustrated an ice-sheet bed with a near-complete landform record but it is much more likely that the landform record will be fragmentary in nature. There will be gaps in the record where landforms are missing, as well as areas where landforms have been modified, overprinted or erased by changes in ice-flow direction or thermal regime.

2. We have assumed that the mapped glacial landforms formed more-or-less synchronously, at or around the ice-sheet maximum. This is not the case in the real world because landform development beneath ice sheets is time-transgressive; that is landforms are created, re-orientated, overprinted or erased as the ice sheet changes its configuration and thermal regime over time (e.g., during ice-sheet growth and during deglaciation). Changes in ice-flow direction would arise from migration of the ice divides (e.g., during ice-margin retreat), or as ice streams switch on and off in response to changing subglacial conditions and changes in ice-sheet basal thermal regime. Landforms relating to ice-sheet inception and ice-sheet growth would probably be destroyed during ice-recession so it is difficult to extract information about these time periods directly from the landform record.

3. Our simple model does not account for wholesale changes in the basal thermal regime of the ice sheet through time; that is, the effects of changes in the location of zones of basal melting and freezing as the ice sheet expands and recedes across the landscape. These changes have the potential to create, re-orientate, overprint or erase existing landforms.

4. Deglaciation is the last in a series of events during the lifetime of an ice sheet and it is likely that most of the landforms would probably date from deglaciation, not from the ice-sheet maximum as we have illustrated. Where the landform record contains features such as recessional moraines, eskers and ice-marginal meltwater channels, these could also be used to reconstruct patterns of ice-recession during deglaciation.

Notwithstanding these limitations, this simple model illustrates some of the landforms we might expect to encounter on a former ice-sheet bed, their spatial distribution and their relationship to one another.

12.5 SUMMARY

Glacial geological inversion models are a conceptual tool used to make palaeoglaciogical reconstructions based on the mapped distribution of landforms and sediments on the beds of former ice sheets. These models provide information about the spatial extent of former ice sheets and how ice-sheet dimensions and behaviour changed through time. Much of the landform information that we see now relates to conditions at the ice-sheet maximum and during deglaciation. Landforms relating to ice-sheet inception and ice-sheet growth are often destroyed during ice-recession so these time periods are underrepresented in the landform record. Palaeoglaciogical reconstructions can be used to test numerical models of former ice sheets and to investigate the wider relationship between glacial erosion and deposition (Box 12.7).

BOX 12.7: ICE-SHEET WIDE PATTERNS OF EROSION AND DEPOSITION

Glacial erosion and glacial deposition are often treated separately, but Kleman *et al.* (2008) provide a neat example of how these two subjects can be combined in a palaeoglaciological reconstruction. These authors examined spatial differences in both glacial erosion, using the patterns of glacial scouring, fjords, lake basins and deep linear erosion, and glacial deposition, using the thicknesses of Quaternary deposits (termed drift: mainly tills) across Fennoscandia. Their ultimate aim was to investigate the relative roles of mountain ice sheets (MIS) and the full-sized Fennoscandian Ice Sheet (FIS) in shaping the landscape. Fjords and deep non-tectonic lakes were used to delineate zones of deep glacial erosion, and relict landscapes were used to mark areas characterised by former frozen-bed conditions. The amount of exposed bedrock in the landscape and drift thickness was used to create a three-fold landscape classification comprised of: (i) thick drift; (ii) intermediate drift thickness; and (iii) absence of drift/ice-scoured zones. The figure below illustrates this. Part A is a map of drift cover thickness and distribution, different types and intensity of glacial erosion, and occurrences of relict landscapes in Fennoscandia. Relict landscapes are defined as consisting of pre-Late Weichselian landforms and deposits (such as boulder fields). Part B shows the location of thick drift cover zones and scouring zones along a NW–SE transect across the area covered by the former Fennoscandian Ice Sheet. Kleman *et al.* (2008) then tried to relate the distribution of these different landscapes to former glaciological regimes. They inferred that the western fjord zone of deep glacial erosion formed beneath both MIS- and FIS-style ice sheets during the entire Quaternary, whereas the eastern (lake) zone of deep glacial erosion was exclusively related to MIS style ice sheets, and formed largely during the early and middle Quaternary. The scouring zones formed under conditions of rapid ice flow towards iceberg-calving margins of FIS-style ice sheets. A centrally placed zone of thick drift could not be explained by deposition under FIS-style ice sheets, so was interpreted to be the combined result of marginal deposition of fluctuating

MIS-style ice sheets, and the inefficiency of later FIS-style ice sheets to evacuate this drift because their core areas were characterised by low ice velocity and frozen-bed conditions.

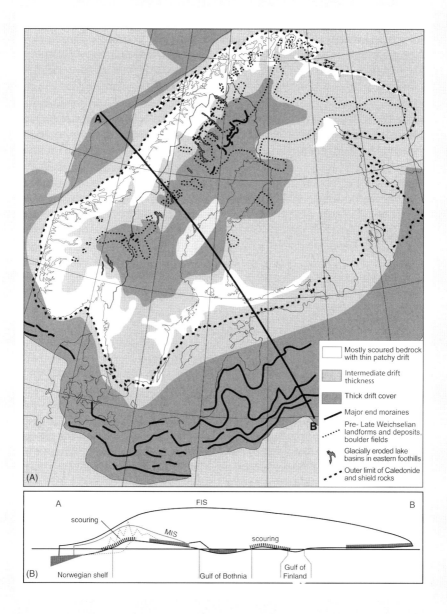

Source: Kleman, J.K., Stroeven, A.P. and Lundqvist, J. (2008) Patterns of Quaternary ice sheet erosion and deposition in Fennoscandia and a theoretical framework for explanation. *Geomorphology*, **97**, 73–90. [Modified from: Kleman *et al.* (2008) *Geomorphology*, **97**, figure 7, p. 80].

Techniques are also being developed that allow us to date palaeoglaciological events more accurately (Box 12.8), which will in turn allow us to improve and refine ice-sheet reconstructions.

BOX 12.8: OPTICALLY STIMULATED LUMINESCENCE (OSL) DATING OF GLACIGENIC SEDIMENTS

Luminescence dating is a method used to determine the age of samples deposited as recently as 30–50 years ago, up to almost a million years. The method is based upon the fact that many naturally occurring minerals, including quartz and most feldspars, act as dosimeters (i.e. they record the amount of ionising radiation that they have been exposed to). This radiation principally comes from the radioactive decay of uranium, thorium and potassium in the sediments surrounding a sample. After being exposed to such radiation, when a sample is heated or exposed to light it emits light. This light is called luminescence. Dating glacial sediments using *Optically Stimulated Luminescence* (OSL) is challenging because many glacial sediments are not exposed to light prior to burial (Duller, 2006). However, with careful selection of sample sites, it has proved possible to date ice-contact glaciofluvial and glaciolacustrine sediments that can be used in palaeoglaciogical reconstructions. Glasser *et al.* (2006) provide an example of this technique from the Rio Bayo Valley, to the northeast of the North Patagonian Icefield, Chile. These authors used single-grain OSL dating to establish that a large outlet glacier was present in the valley until at least 9.7 ± 0.7 ka. They also used cosmogenic nuclide exposure age dating (see Box 6.10) of erratic boulders in the valley as an independent test of this age estimate. They found a good agreement between the ages of former glacier extent using these two independent dating techniques (OSL and cosmogenic nuclide exposure age dating).

Sources: Benn, D.I. and Owen, L.A. (2002) Himalayan glacial sedimentary environments: A framework for reconstructing and dating the former extent of glaciers in high mountains. *Quaternary International*, **97/8**, 3–25. Duller, G.A.T. (2006) Single grain optical dating of glacigenic sediments. *Quaternary Geochronology*, **1**, 296–304. Glasser, N.F., Harrison, S., Ivy-Ochs, S. *et al.* (2006) Evidence from the Rio Bayo valley on the extent of the North Patagonian icefield during the late Pleistocene-Holocene transition. *Quaternary Research*, **65**, 70–7.

SUGGESTED READING

Clark (1997) outlines how remote sensing and GIS are used in reconstructing former ice sheets from geomorphological and geological evidence. Jansson and Glasser (2005a) explain how satellite imagery and Digital Elevation Models are used to map glacial lineations to make ice-mass reconstructions. Stokes and Clark (2001) review geomorphological evidence for palaeo-ice streams, and Stokes and Clark (2002) describe ice stream shear-margin moraines. Patterson (1998), Evans *et al.* (2008) and De Angelis and Kleman (2007, 2008) describe glacial geomorphological and sedi-mentological evidence for former fast glacier flow (ice streaming) in various parts of the Laurentide Ice Sheet. Marine palaeo-ice streams and their geomorphological and sedimentological signature are described by Dowdeswell *et al.* (2004, 2007), Ó Cofaigh *et al.* (2005, 2007), Smith *et al.* (2007) and Ottesen *et al.* (2005). Anderson (1999) considers the contribution from the Antarctic marine geological record. Greenwood *et al.* (2007) outline how glacial meltwater channels can be used in ice-sheet reconstructions. Goodfellow (2007) reviews evidence for the preservation of relict surfaces in glaciated areas, and Kleman and Glasser (2007) outline the subglacial thermal organisation of former ice sheets. Hättestrand and Kleman (1999) outline theories of ribbed moraine development and their palaeoglaciologi-cal significance. Kjær *et al.* (2003) show how the erratic content in tills can be used to reconstruct the flow patterns and dynamic behaviour of the southwestern margin of the last Scandinavian Ice Sheet. Iverson (2004) provides a neat example of how quantitative estimates of glacier sliding velocities can be inferred from sedimentary evidence.

A number of papers combine landform and sedimentary evidence into palaeo-glaciologcal reconstructions for former ice sheets. These include the Laurentide Ice Sheet (Sugden, 1977; Dyke and Prest, 1987; Boulton and Clark, 1990; Clark *et al.*, 1996), the European Ice Sheet/Fennoscandian Ice Sheet (Kleman *et al.* 1997; Boulton *et al.*, 2001), the British–Irish Ice Sheet (Clark *et al.*, 2004, 2006; Evans *et al.*, 2005; Jansson and Glasser, 2005b; Roberts *et al.*, 2007), the Patagonian Ice Sheet (Glasser *et al.*, 2008), the Svalbard–Barents Sea Ice Sheet (Siegert and Dowdeswell, 1995) and the Eurasian Ice Sheet (Svendsen *et al.*, 2004). Finally, at the valley-glacier scale, Carr and Coleman (2007) and Harrison (2005) provide reviews of the methods used to reconstruct Pleistocene valley glaciers and their former ELAs.

Anderson, J.B. (1999) *Antarctic Marine Geology*. Cambridge University Press.

Boulton, G.S. and Clark, C.D. (1990) A highly mobile Laurentide ice sheet revealed by satellite images of glacial lineations. *Nature*, **346**, 813–917.

Boulton, G.S., Dongelmans, P., Punkari, M. and Broadgate, M. (2001) Palaeoglaciology of an ice sheet through a glacial cycle: the European ice sheet through the Weichselian. *Quaternary Science Reviews*, **20**, 591–625.

Carr, S. and Coleman, C. (2007) An improved technique for the reconstruction of former glacier mass-balance and dynamics. *Geomorphology*, **92**, 76–90.

Clark, C.D. (1993) Mega-scale glacial lineations and cross-cutting ice-flow landforms. *Earth Surface Processes and Landforms*, **18**, 1–29.

Clark, C.D. (1997) Reconstructing the evolutionary dynamics of former ice sheets using multi-temporal evidence, remote sensing and GIS. *Quaternary Science Reviews*, **16**, 1067–92.

Clark, C.D., Evans, D.J.A., Khatwa, A., *et al.* (2004) Map and GIS database of glacial landforms and features related to the last British ice sheet. *Boreas*, **33**, 359–75.

Clark, C.D., Greenwood, S.L. and Evans, D.J.A. (2006) Palaeo-glaciology of the last British–Irish Ice Sheet: Challenges and some recent developments, in *Glacier Science and Environmental Change* (ed. P. Knight), Blackwell, Oxford, pp. 248–64.

Clark, P.U., Licciardi, J.M., MacAyeal, D.R. and Jenson, J.W. (1996) Numerical reconstruction of a soft-bedded Laurentide ice sheet during the last glacial maximum. *Geology*, **24**, 679–82.

De Angelis, H. and Kleman, J. (2007) Palaeo-ice streams in the Foxe/Baffin sector of the Laurentide ice sheet. *Quaternary Science Reviews*, **26**, 1313–31.

DeAngelis, H. and Kleman, J. (2008) Palaeo-ice-stream onsets: Examples from the north-eastern Laurentide ice sheet. *Earth Surface Processes and Landforms*, **33**, 560–72.

Dowdeswell, J.A., O'Cofaigh, C. and Pudsey, C.J. (2004) Thickness and extent of subglacial till layer beneath an Antarctic paleo-ice stream. *Geology*, **32**, 13–16.

Dowdeswell, J.A., Ottesen, D., Rise, L. and Craig, J. (2007) Identification and preservation of landforms diagnostic of past ice-sheet activity on continental shelves from three-dimensional seismic evidence. *Geology*, **35**, 359–62.

Dyke, A.S. and Prest, V.K. (1987) Late Wisconsinan and Holocene history of the Laurentide ice sheet. *Géographie Physique Et Quaternaire*, **41**, 237–63.

Evans, D.J.A., Clark, C.D. and Mitchell, W.A. (2005) The last British ice sheet: a review of the evidence utilised in the compilation of the glacial map of Britain. *Earth Science Reviews*, **70**, 253–312.

Evans, D.J.A., Clark, C.D. and Rea, B.R. (2008) Landform and sediment imprints of fast glacier flow in the southwest Laurentide ice sheet. *Journal of Quaternary Science*, **23**, 249–72.

Glasser, N.F., Harrison, S., Jansson, K. and Kleman, J. (2008) The glacial geomorphology and pleistocene history of southern South America between 38°S and 56°S. *Quaternary Science Reviews*, **27**, 365–90.

Goodfellow, B.W. (2007) Relict non-glacial surfaces in formerly glaciated landscapes. *Earth Science Reviews*, **80**, 47–73.

Greenwood, S.L., Clark, C.D. and Hughes, A.L.C. (2007) Formalising an inversion methodology for reconstructing ice-sheet retreat patterns from meltwater channels: Application to the British ice sheet. *Journal of Quaternary Science*, **22**, 637–45.

Hättestrand, C. and Kleman, J. (1999) Ribbed moraine formation. *Quaternary Science Reviews*, **18**, 43–61.

Harrison, S.P. (2005) Snowlines at the last glacial maximum and tropical cooling. *Quaternary International*, **138/139**, 5–7.

Iverson, N.R. (2004) Estimating the sliding velocity of a Pleistocene ice sheet from plowing structures in the geologic record. *Journal of Geophysical Research*, **109**, F04006. (doi 10.1029/2004JF000132.)

Kjær, K.H., Houmark-Nielsen, M. and Richardt, N. (2003) Ice-flow patterns and dispersal of erratics at the southwestern margin of the last Scandinavian ice sheet: Signature of palaeo-ice streams. *Boreas*, **32**, 130–48.

Jansson, K.N. and Glasser, N.F. (2005a) Palaeoglaciology of the Welsh sector of the British–Irish ice sheet. *Journal of the Geological Society, London*, **162**, 25–37.

Jansson, K.N. and Glasser, N.F. (2005b) Using Landsat 7 ETM+ imagery and Digital Terrain Models for mapping glacial lineaments on former ice sheet beds. *International Journal of Remote Sensing*, **26**, 3931–41.

Kleman, J. and Glasser, N.F. (2007) The subglacial thermal organisation (STO) of ice sheets. *Quaternary Science Reviews*, **26**, 585–97.

Kleman, J., Hättestrand, C., Borgström, I. and Stroeven, A. (1997) Fennoscandian palaeoglaciology reconstructed using a glacial geological inversion model. *Journal of Glaciology*, **144**, 283–99.

O'Cofaigh, C., Dowdeswell, J.A., Allen, C.S., *et al.* (2005) Flow dynamics and till genesis associated with a marine-based Antarctic palaeo-ice stream. *Quaternary Science Reviews*, **24**, 709–40.

O'Cofaigh, C., Evans, J., Dowdeswell, J.A. and Larter, R.D. (2007) Till characteristics, genesis and transport beneath Antarctic paleo-ice streams. *Journal of Geophysical Research*, **112**, F03006, doi:10.1029/2006JF000606.

Ottesen, D., Dowdeswell, J.A. and Rise, L. (2005) Submarine landforms and the reconstruction of fast-flowing ice streams within a large Quaternary ice sheet: the 2500-km-long Norwegian–Svalbard margin (57°–80°N). *Geological Society of America Bulletin*, **117**, 1033–50.

Patterson, C.J. (1998) Laurentide glacial landscapes: The role of ice streams. *Geology*, **26**, 643–6.

Roberts, D.H., Dackombe, R.V. and Thomas, G.S.P. (2007) Palaeo-ice streaming in the central sector of the British-Irish ice sheet during the last glacial maximum: evidence from the northern Irish sea basin. *Boreas*, **36**, 115–29.

Siegert, M.J. and Dowdeswell, J.A. (1995) Numerical modeling of the Late Weichselian Svalbard–Barents Sea ice sheet. *Quaternary Research*, **43**, 1–13.

Smith, A.M., Murray, T., Nicholls, K.W., *et al.* (2007) Rapid erosion, drumlin formation, and changing hydrology beneath an Antarctic ice stream. *Geology*, **35**, 127–30.

Stokes, C.R. and Clark, C.D. (1999) Geomorphological criteria for identifying Pleistocene ice streams. *Annals of Glaciology*, **28**, 67–74.

Stokes, C.R. and Clark, C.D. (2001) Palaeo-ice streams. *Quaternary Science Reviews*, **20**, 1437–57.

Stokes, C.R. and Clark, C.D. (2002) Ice stream shear margin moraines. *Earth Surface Processes and Landforms*, **27**, 547–58.

Sugden, D.E. (1977) Reconstruction of the morphology, dynamics, and thermal characteristics of the Laurentide ice sheet at its maximum. *Arctic and Alpine Research*, **9**, 21–47.

Svendsen, J.I., Alexanderson, H., Astakhov, V.I., *et al.*, (2004) Late Quaternary ice sheet history of northern Eurasia. *Quaternary Science Reviews*, **23**, 1229–71.

Index

Note: Entries in bold are **key** entries.

potholes   146
sichelwannen   138, 146–7
Piedmont lobes   74
Pinedale Ice Mass *see* North America
Pine Island Glacier *see* Antarctica
Pinning points   **73**, 322–3, 336
Ploughing clasts   209
Plough marks   335, 341–344
Plucking (quarrying)   109, 117, 133, 150–3, 261
Polar ice *see* Cold ice under Basal Thermal Regime
Polar plateau   8
Polythermal glaciers *see* Mixed regime (polythermal) glaciers
Pore water pressure   54, 215, 256, **278**, 281
Positive feedback   75, **153**, 165, 172–3
Postglacial period   5
Post-sedimentary   239–40
Precambrian   202–3
Preglacial landscapes   126, 367
Pressure melting point   19, 30, 52, **59**
Pressure release   119
Prince George, British Columbia *see* Canada
Principle of superposition   209–10
Proglacial lake   20, 23, 28
Proglacial stream   29
Push moraines   248–58
    composite   248
    seasonal/annual   248–51

Radioactive decay   372
Rain-out   **307**, 313, **316**, 318–19, 321, 342–3
R-channels (Röthlisberger)   88
Regelation   192
Regelation ice   **52**, 124, 186
Regelation slip *see* Basal sliding
Relief   158, 292
Rheology   **55**, 117, 217, 260, 278, 283, 286
Roche moutonnées   110, 116, 120, 148–51
    evolution   151–2
    geological structure   150, 152
    giant roche moutonnée   110, 116, 120, 149–52, 160
    morphology   150–2
    significance   148
Rock basins   148, 153, 157, 159–60, 169, 172, 179
Rock drumlins   149
Rock grooves and basins   147
Rogen   248, 269, **270**, 286, 299, 357
    rogen (ribbed) moraines   275, **283**, 284
Ross Ice Shelf *see* Antarctica

Salpausselkä Moraines (Finland)   331, 332
Sandur (outwash plain)   291–3, 355, 357
San Rafel Glacier *see* Patagonia
Saskatchewan Glacier *see* Canada
Scandinavian Ice Sheet   125, 332, 365
Scarborough Bluffs *see* Canada
Scotland   89, 166, 169, 173, 176, 179, 204, 286, 294, 298, 300, 363–4
    Cairngorms   176
    Carstairs Kames   284, 298
    Glen Dee   173
    Isle of Mull   89
    Loch Etive   294–5
    Scottish Highlands   74, 171–2, 265, 284
Scott Glacier   241
Sea level change   7, 10, 13, 16–18, **41–2**, 49, 75, 83, 341, 344
Sefstrombreen *see* Svalbard
Seward Glacier *see* Alaska
Shear margins   8
Shear planes   187, 198, 209, 225, 228–9
Shear stress   50–1, 70, 75–6, 79, 122, 213, 215, 230, 365
    basal sheer stress   47–9
    in glacial reconstructions   49–50
Sheet joints   119
Shreve's equation   95
Sidestrand *see* England
Siple Coast *see* Antarctica
Skeidararjökull *see* Iceland
Sléttjökull *see* Iceland
Smudges   209, 225
Snowball Earth   1, 3
Snowdon *see* Wales
Snowline *see* Equilibrium line
Sognefjord (Norway)   126
South Asian monsoon   22
Southern Alps *see* New Zealand
Southern Ocean   156–7
South Shetland Islands *see* Antarctica
Sprag marks   342
Spring event   17
Squeeze moraines   22
Stacked till sheets   261
Stagnant ice   21–2, 82, 300, 305
Sticky spots   122, 215
Stoss and lee forms   116, 141, 147–8, **150–3**, 158, 179
    giant stoss and lee forms   157–8, **173**, 174
Strain markers   54
Streamlined bedrock features   147–50
Streamlined hills   149, 173